# COPYRIGHT, COMPETITION AND INDUSTRIAL DESIGN

*To the memory of Professor W A Wilson*

THE DAVID HUME INSTITUTE

Hume Papers on Public Policy
Volume 3 No 2 Summer 1995

# COPYRIGHT, COMPETITION AND INDUSTRIAL DESIGN
## Second Edition

Hector L. MacQueen

EDINBURGH UNIVERSITY PRESS

© David Hume Institute 1995

Edinburgh University Press
22 George Square, Edinburgh

Typeset in Times New Roman by WestKey Limited, Falmouth, Cornwall and printed and bound in Great Britain by Page Bros. Limited, Norwich

A CIP record for this book is available from the British Library

ISBN 0 7486 0733 1

# Contents

Foreword — vii

1. Introductory — 1
2. The Rise and Fall of Design Copyright — 25
3. The Road to Reform — 49
4. The Copyright Designs and Patents Act 1988 — 63
5. The European Future — 87

Notes — 104
Bibliography — 105
Index — 110

# Author

**Hector L MacQueen** is Director of The David Hume Institute and Professor of Private Law in the University of Edinburgh.

# Foreword

The first edition of this work appeared in 1989 and was essentially an explanation and critique of the provisions on the protection of industrial designs contained in the Copyright Designs and Patents Act 1988. A second edition has been prompted by the appearance of European Commission proposals for the protection of designs, and the sometimes controversial case law on the British legislation since 1989. The technical discussion is set against a background of the general considerations currently affecting intellectual property law, in particular its relationship with competition law. It is argued that in exceptional circumstances it is appropriate to bring competition considerations to bear upon intellectual property rights, since the principal economic justification for such rights is the public interest in competition by innovation and creative activity. Where intellectual property in fact stifles rather than facilitates such activity, control is legitimate and necessary. Further, the scope of intellectual property and its role in a market economy is always in need of supervision and review, to ensure that its goals continue to be achieved.

It is even more than usual incumbent upon the Director of The David Hume Institute to disclaim any support from the Institute for the author's views.

Hector L MacQueen
Director
The David Hume Institute

# Acknowledgements

My debts in revising this work are as great as those incurred in the first writing of it. In no particular order of levels of indebtedness, I wish to record my thanks to John Adams, Judge David Edward, Brian Main, Kathy Mountain, Alan Peacock, and Richard Tompson, various colleagues on the Intellectual Property Working Party of the Law Society of Scotland, and the students of the Faculty of Law in Edinburgh University who have been members of my honours class in intellectual property between 1993 and 1995 or who have put up with my supervision of their doctoral and masters research. I greatly benefited from my attendance at the ICARE Conference on the economics of intellectual property rights held in Venice in October 1994. William Windram once again helped with proofs. Errors of fact, law and opinion are all my own.

The first edition was dedicated to the late Sir Thomas Smith who had died little more than a year before its publication. It is with sorrow but also great gratitude that I dedicate this second edition to the memory of another professor of law at Edinburgh taken too early from his friends, Bill Wilson (1929-1994). He introduced me as a student to the subject of intellectual property, and as a colleague encouraged my interest in it. It is a great sadness that I cannot simply thank him here for his help in completing this book, as I did in the first edition.

HLM

# Chapter One

## Introductory

### Some Historical Perspectives

David Hume the philosopher, the eponym of The David Hume Institute, does not seem to have written very much on the subject of copyright even though it was a controversial subject in his own time and he himself was at one stage caught up around the fringes of the debate. In his well-known biography of Hume, Ernest Mossner narrates how the initial sales of *The History of England*, published in six volumes between 1754 and 1762, were retarded by the 'Conspiracy of the London Booksellers'. The *History* was first published in Edinburgh at a time when the London publishers, who had hitherto dominated the market, were using all means in their power to restrict competition emanating from Edinburgh and elsewhere. This involved "one of the most successful propaganda campaigns in the history of the restraint of trade". Mossner explains:

> Using every means at hand to reduce the success of books which they did not own, they developed a highly efficient communication system. It extended to provincial booksellers, to whom riders brought news of books – which to push, which to pass by. It often included direct threat of trade retaliation. It sometimes involved an offer to buy entire stocks (for which 'correct editions' would be substituted). It nearly always relied upon a determined and damaging criticism of the books to be curtailed.

In the end a campaign of this kind compelled Hume to abandon his Edinburgh publisher, Gavin Hamilton, and to transfer to Andrew Millar of London, who produced the remaining volumes of the *History*. It went on to become an immense success (Mossner 1980: 312-16, quotations at 313-14).

Propaganda and threats of trade retaliation were not the only means by which the London publishers of the eighteenth century protected themselves from competition. They also made use of the then infant law of copyright. The Copyright Act 1709 had created a system applicable throughout Britain whereby upon registration of a book in Stationers Hall, London, an author and his assigns gained for a period of fourteen years from the date of first publication "the sole right and liberty of printing" the book. If the author was still alive at the end of the fourteen years, the right continued for a second fourteen-year term. In fact it was the London booksellers and publishers who

## 2 COPYRIGHT, COMPETITION AND INDUSTRIAL DESIGN

benefited most from the Act, with their ready access to Stationers Hall. But towards the middle of the eighteenth century they began to run into difficulties with competition from Scotland, one aspect of which was the republication by the Scottish booksellers of works in which the statutory copyright of the London booksellers had expired. In a series of cases in both English and Scottish courts the London booksellers argued for the existence of a perpetual copyright at common law, regardless of the expiry of the statutory period of protection. Hume's Scottish publisher, Gavin Hamilton, had been involved in such a case in 1748, *Midwinter v Hamilton* (1748) Mor 8295. Where successful, as for example in the English decision of *Millar v Taylor* (1769) 4 Burr 2303, the argument enabled the London booksellers to close out the competition from Scotland and elsewhere. In England common law copyright was only finally refuted in the great case of *Donaldson v Beckett* (1774) 2 Bro PC 129; in Scotland, it had already met its demise the previous year when the Court of Session decided the case of *Hinton v Donaldson* (1773) Mor 8307. (See further Feather 1988: 77-83; Tompson 1992; Rose 1993.) Hume thought the cases to have been correctly decided as a matter of the construction of the 1709 Act but supported the booksellers in their subsequent pursuit of a reform in the law to reinstate their rights (Greig 1969: ii, 286-8; *Parliamentary History*: xvii, cols 1098, 1108, 1400). Despite the failure of this campaign to make much headway, Hume's *History* seems still to have been earning money for its publishers in 1805, nearly 30 years after his death (Mossner and Ransom 1950: 180), perhaps because Hume himself encouraged them to put out amended editions of his work in order to create new copyrights under the 1709 Act (Greig 1969: ii, 286-8).

The interest of this history for present purposes is that it shows the rights conferred by copyright being used to restrain trade competition. It is of the essence of copyright that it confers a right upon a person to stop others copying his work without his consent. Obtaining that consent will normally cost money. Hence, for example, a publisher must pay an author a fee or royalties for the privilege of publishing his work. It is mildly paradoxical, perhaps, that generally to maximise his benefit from copyright the author actually wants as many copies made as possible; but without copyright and the ability to control the reproduction of his work there would be no return to him, no matter how popular or successful it had been. In the words of an introductory textbook on the subject, "the Copyright Acts... were meant for the protection of authors, artists and composers and to provide a legal foundation for the innumerable transactions by which authors, artists and composers are paid for their work" (Jacob and Alexander 1993: 125). Such transactions may take the form of assignations – that is, outright transfer of the copyright to another – or licences, which are permissions to make copies, perhaps of a particular kind, for a defined period of time, and in a particular geographical area, but under which ownership of the copyright is retained while the licensee engages in his permitted activity. More than one licence may be granted, enabling the copyright owner to exploit different markets for his work through those best equipped to deal in and with those various markets. For example, a book may be published in several countries; there may also be serialisation rights, dramatisation rights, film rights, and broadcasting rights, again in several countries, to be considered as possible methods of exploitation. Here licensing comes into its own and gives the copyright owner considerable power to control the market in his work.

The strategy adopted by Sir Walter Scott and his publisher Robert Cadell to pay off the former's creditors following his insolvency in 1826, which was similar to that suggested by Hume in 1774, demonstrates the economic value of copyright to authors and publishers. Scott had previously sold the copyrights in the Waverley Novels to his partner Archibald Constable, the publisher of the series. Constable crashed along with Scott but his trustees sold the Waverley copyrights to Scott's trustees in partnership with Cadell. Scott and Cadell then embarked upon a new annotated edition of the novels – the Magnum edition – free from any alternative claim. Morover a fresh copyright would be established by Scott's introductions, textual revisions and annotations, which removed the threat that as the copyrights in the first editions expired other publishers would step in with rival republications. Accordingly all possible profit from Scott's work would continue to come to the author and his publisher over a new copyright term, and the creditors' prospect of repayment in full would be greatly enhanced. Cadell was careful to have bound the interleaved volumes on which Scott worked, because "it provided evidence of the continuation of the copyright in the Waverley Novels" and preservation was essential; these volumes, surely of more than passing interest to the historian of copyright as well as the literary critic, are now located in the National Library of Scotland in Edinburgh (Millgate 1987: 5-8, 47, 51, 53, 64; Millgate 1986: 7-9, 14-15). The success of Cadell's strategy can be seen in the case of *Black v Murray* (1870) 9 M 341, decided just one year before the Scott Centenary Exhibition in 1871. There the copyright in the later editions of Scott's work, in particular the annotations, was successfully upheld in the Court of Session by the publishers A & C Black, to whom Cadell's family had sold the rights after his death in 1849.

Although eighteenth- and nineteenth-century exploitation of copyright in the market was much less sophisticated than is possible today, the activities of the London booksellers and of Scott and Cadell show that already its market power was keenly appreciated (see also Feather 1988: 170). Lawyers also understood this well, which led to fierce juristic debate, in particular in the cases of *Midwinter v Hamilton* (1748) and *Hinton v Donaldson* (1773), as to whether or not copyright was the recognition of a form of property arising from the act of creation, or merely a statutory monopoly founded in the public interest. In Scotland the firm view came to be that copyright was not a right of property but a form of statutory monopoly, restricting for reasons of public policy what would otherwise be the natural liberties of mankind. That being so, copyright was not attended with the consequences of property ownership in general; rather it was, like all monopolies, to be approached narrowly to avoid the potential evil consequences of monopoly. Lord Kames' opinion rejecting common law copyright in *Hinton v Donaldson*, as reported by James Boswell (1774: 18-21), shows his awareness of the possible consequences of allowing too great scope for the exercise of monopoly power by extending the law:

The act of Queen Anne is contrived with great judgment, not only for the benefit of authors, but for the benefit of learning in general. It excites men of genius to exert their talents for composition; and it multiplies books both of instruction and amusement. And when, upon expiration of the monopoly, the commerce of these books is laid open to all, their cheapness, from a concurrence of many editors, is singularly beneficial to the public. Attend, on the other hand, to the consequences of a perpetual monopoly. Like all other monopolies, it will unavoidably raise the price of good books beyond the reach of ordinary readers. They will be sold like so many valuable pictures, the sale will be confined to a few learned men who have money to spare, and to a few rich men who buy out of vanity as they buy a diamond or a fine coat. The commerce of books will be in a worse state than before printing was invented: at that time, manuscript copies might be multiplied at pleasure; but even manuscript copies would be unlawful if there were a perpetual monopoly. Fashions at the same time, are variable; and books, even the most splendid, would wear out of fashion with men of opulence, and be despised as antiquated furniture. The commerce of books would of course be at an end; for even with respect to men of taste, their number is so small, as of themselves not to afford encouragement for the most frugal edition. Thus booksellers, by grasping too much, would lose their trade altogether; and men of genius would be quite discouraged from writing, as no price can be afforded for an unfashionable commodity. In a word, I have no difficulty to maintain that a perpetual monopoly of books would prove more destructive to learning and even to authors, than a second irruption of Goths and Vandals. And hence with assurance I infer, that a perpetual monopoly is not a branch of the common law or of the law of nature. God planted that law in our hearts for the good of society; and it is too wisely contrived to be in any case productive of mischief.

Our booksellers, it is true, aiming at present profit, may not think themselves much concerned about futurity. But it belongs to judges to look forward; and it deserves to be duly pondered whether the interest of literature in general ought to be sacrificed to the pecuniary interest of a few individuals. The greatest profit to them ought to be rejected, unless the monopoly be founded in common law beyond all objection: the most sanguine partizan of the booksellers will not pretend this to be the case. At the same time, it will be found, upon the strictest examination, that the profit of such a monopoly would not rise much above what is afforded by the statute. There are not many books that have so long a run as fourteen years; and the success of books upon the first publication is so uncertain, that a bookseller will give very little more for a perpetuity, than for the temporary privilege bestowed by the statute. This was foreseen by the legislature; and the privilege was wisely confined to fourteen years; a sufficient encouragement to men of genius without hurting the public interest. The best authors write for fame: the more diffused their works are, the more joy they have. The monopoly then is useful only to those who write for money or for bread, who are not always of the most dignified sort. Such writers will gain very little by the monopoly; and whatever they may gain at present, the profits will not be of long endurance; a monopoly would put a final end to the commerce of books in a few generations. And therefore, I am for dismissing this process as contrary to law, as ruinous to the public interest, and as prohibited by statute.

There are many interesting points about this passage of Kames, in particular his stress upon the public interest as the factor which both justifies the monopoly of copyright and limits it. The monopoly provides an incentive and reward for activity in the public interest; but if it is made too strong then it will tend to defeat the end for which it is granted. The suggestion that the best authors write for fame rather than money of course brings to mind Dr Johnson's well-known observation, made only a few years later and also recorded by the ubiquitous Boswell, that "no man but a blockhead ever wrote except for money". Boswell, it may be added, thought Johnson mistaken (Hill and Powell 1934-64: iii, 19-20).

An emphasis similar to that of Kames, on the balance to be struck between the public interest in an author's right to reward and the public interest in having access to his work, can be detected in the works of Adam Smith. In his lectures on jurisprudence, delivered at the University of Glasgow in the early 1760s, Smith dealt with copyright (and patents) in his discussion of exclusive privileges as real rights. In general, Smith argued, exclusive privileges "are greatly prejudicial to society" as tending toward monopolies; but copyright was "harmless". It could be justified "as an encouragement to the labours of learned men" (Meek *et al* 1978: 83). A passage in *The Wealth of Nations* (1776) suggests that he may have thought this encouragement desirable. After referring to "that unprosperous race of men commonly called men of letters", he comments that before the invention of printing the only remunerative occupation for them was teaching and "a scholar and a beggar seem to have been terms very nearly synonymous (*sic*)" (Todd 1976: i, 148-49). In the lectures Smith stressed the balance which could be struck by virtue of copyright between the worth of a work and the author's reward. "And this," he continued, "is perhaps as well adapted to the real value of the work as any other, for if the book be a valuable one the demand for it in that time will probably be a considerable addition to his fortune. But if it is of no value the advantage he can reap from it will be very small" (Meek *et al* 1978: 83).

The general view of copyright in its early days seems to have been a restrictive one, strongly led by the economic consideration that if the law became too protective it would lead to the creation of undesirable monopolies. There was a slow departure from this position in the course of the nineteenth century as copyright was gradually extended in scope and in the length of time for which it lasted. The emphasis in the philosophy of the law fell increasingly on the author's rights of property in his work. Something of this seems to underlie Samuel Johnson's remark to Boswell, that "there seems to be in authors a stronger right of property than that by occupancy; a metaphysical right, as it were, of creation, which should from its nature be perpetual" (Hill and Powell 1934-50: ii, 259), as well as his criticism of the Court of Session's steadfast refusal to recognise any copyright apart from that provided by the Act of 1709. Boswell narrates:

> I read to him my notes of the Opinions of our Judges upon the question of Literary Property. He did not like them; and said, "they make me think of your Judges not with that respect which I should wish to do". To the argument of one of them, that there can be no property in blasphemy or nonsense, he answered, "then your rotten sheep are mine! – By that rule, when a man's house falls into decay, he must lose it" (Hill and Powell, 1934-64: v, 50).

## 6 COPYRIGHT, COMPETITION AND INDUSTRIAL DESIGN

Johnson did however recognise that the author's perpetual property right had to be curtailed in the public interest and his overall view was not so far from that of Kames and Smith:

> [T]he consent of nations is against it; and indeed reason and the interest of learning are against it; for were it to be perpetual, no book, however useful, could be universally diffused amongst mankind, should the proprietor take it into his head to restrain its circulation. No book could have the advantage of being edited with notes, however necessary to its elucidation, should the proprietor perversely oppose it. For the general good of the world, therefore, whatever valuable book has once been created by an author, and issued out by him, should be understood as no longer in his power, but as belonging to the public; at the same time the author is entitled to an adequate reward. This he should have by an exclusive right to his work for a considerable number of years (Hill and Powell 1934-64: ii, 259).

The idea that property arises from the act of creation so that copying becomes a form of theft is an old one. It can be found in the story of St Columba surreptitiously copying the manuscript psalter made by Finnian of Moville, for which judgment was given against him by Diarmaid the king of Tara. Diarmaid is said to have remarked "to every cow its calf, to every book its little book", a judgment which Columba condemned as "crooked" (Kelly 1988: 239-40). David Hume's nephew, also David Hume, professor of Scots Law at Edinburgh University from 1786 to 1822, rationalised the matter thus in his lectures on the topic:

> The Book has a special character – an individual character of its own, which it owes entirely to the particular author, and never could have received from any one but him. Had it not been for him, this particular book, such as it is in point of matter, arrangement, expression and so forth, compleatly distinguishable from all other books on the same subject, never could have existed (Paton 1940-58: iv, 64).

It is interesting to note that although Hume dealt with copyright under the heading 'Exclusive Privilege' his discussion nowhere refers to the issues of balancing the rights of authors against those of the public at large which had preoccupied earlier writers. He treats the matter of common law copyright as settled in the negative by authority but hints that considerations of justice and equity may lie the other way, that is, in favour of the author (Paton 1940-58: iv, 64-72). Hume's successor in the Edinburgh chair, George Joseph Bell, followed him in emphasising the property right of the author – "Of all things, the produce of a man's intellectual labour is most peculiarly distinguishable as his own" – and argued that "the statutes on which Copyright now rests, are intended not so much to create a right, as to protect it against invasion" (MacLaren 1870: i, 103).

From the late eighteenth century onward an increasing emphasis on the author's rights is detectable in the development of the law (Feather 1988: 171), although this never went as far in Britain as it was to do in France and Germany, where *droit d'auteur* and *urheberrecht* came to be justified principally as a protection for the expressions of the author's personality (Davies 1994: 73-134). Use of the word "plagiarism" – derived from the Roman law term for the kidnapping of a child, *plagium* – to describe the activities of the infringer of copyright reflected the idea that copyright essentially began with the author. In Scotland a series of cases presented difficult questions about rights in material unpublished in its author's lifetime, notably the letters of the Earl of Chesterfield and Robert Burns (*Dodsley v McFarquhar* (1775) Mor 8308, and Appendix, 'Literary Property'). Although so far as copyright was concerned the decisions in these cases adhered to the line established by *Hinton v Donaldson*, the Court of Session did hold that the recipient of letters did not thereby acquire the right to publish them. Such an approach was also to underlie the decision almost a century later in *Caird v Sime* (1887) 14 R (HL) 37, where a professor of moral philosophy at Glasgow University was held entitled to stop publication by others of his lectures. Professor Bell dealt with the matter of unpublished letters as an aspect of copyright in his *Commentaries* (MacLaren 1870: i, 111-14). Although English judges showed themselves more sympathetic than their Scottish brethren to some notion of copyright at common law in unpublished material, it was not until 1911 that the Copyright Act finally removed the requirement of registration in Stationers Hall for published matter. From that Act copyright in general came into existence with the work. Even publication ceased to be a significant event for these purposes: the copyright conferred by the Act existed whether or not the work was published.

This development was linked with extensions of the period for which copyright endured, which was ever more firmly tied to the lifetime of the author. In 1814 legislation extended copyright protection to twenty-eight years from publication or the author's lifetime, whichever was the longer. In the 1830s there were attempts to have the term extended to the author's lifetime plus a period of sixty years, supported by Thomas Carlyle (1839) amongst others on the ground that the return for the author's labour should enure to the benefit of those dear to him when they most needed it. The campaign foundered on opposition most neatly encapsulated in Lord Macaulay's (1853: i, 292) famous definition of copyright as "a tax on readers for the purpose of giving a bounty to authors". Nonetheless, by an Act of 1842 (Talfourd's Act, so named after its sponsor, Serjeant Talfourd, who was acting amongst other things in the interests of the family of the poet William Wordsworth[1]), the term became one of forty-two years from publication or the author's lifetime plus seven years, whichever was the longest – a further move in the direction of linking the right firmly to the author. It was this Act which kept the copyright in Scott's Magnum edition alive to be successfully defended in *Black v Murray* in 1870. The 1911 Act completed the switch from a defined period by laying down that in general where a work was published the copyright would subsist for the author's lifetime plus fifty years. A work would have a potentially perpetual copyright while it remained unpublished; if publication came after the death of the author, the copyright would subsist for fifty years from that date.

Copyright was also being extended as to subject matter. The court decisions that there was no copyright at common law meant that this extension occurred principally through statute. Engravings had had protection since 1734. Legislation in 1798 and 1814 gave copyright to sculptures and in 1862 the Fine Arts

Copyright Act did the same for paintings, drawings and photographs. The concept of copyright as a right to stop the making of copies was also developed by legislation in relation to dramatic and musical works, which enabled the right owner to prevent not only the multiplication of copies but also unauthorised performances of the work. Again this reflects the strengthening of the notion of an author's proprietorial rights in his work. Although the 1911 Act pulled the law together into a single quasi-code, its structure reflected something of the previous piecemeal development. Thus for instance the terms of copyright were not uniform: for artistic works the term was generally to be author's lifetime plus fifty years, without any reference to publication whatsoever. The Act also recognised the development of technology by conferring copyright on sound recordings.

What the nineteenth-century cases made clear was that the law was not concerned with the literary or artistic quality or aesthetic merit of work before copyright could be established. In *Alexander v Mackenzie* (1847) 9 D 748, for example, the Court of Session held that there was copyright in a collection of conveyancing styles. Other nineteenth-century Scottish cases held there to be copyright in a list of imports and exports of goods called the 'Clyde Bill of Entry', in a trade price list and in a railway timetable (*Maclean v Moody* (1858) 20 D 1154; *Harpers v Barry Henry* (1892) 20 R 133; *Leslie v Young* (1893) 21 R (HL) 57). The only requirement was that a work should be 'original' – that is to say, the product of the author's own independent skill and labour. But this did not mean that the skill and labour should produce new thinking or fresh ideas; it was enough that what was produced was an independently formed way of expressing what was already in circulation. There was no copyright in ideas. Legislation reinforced this in some respects: the 1911 Act stated that certain artistic works should have copyright 'irrespective of artistic quality'. The approach in deciding whether or not a work should have copyright was most neatly summarised in an English case concerning the rights in examination papers, *University of London Press v University Tutorial Press* [1916] 2 Ch 601 at 610, where Peterson J remarked that "what is worth copying is worth protecting".

## Expanding Copyright to Meet Technological Development

The development of copyright since the 1911 Act has confirmed some of these expansive trends. Author rights have been strengthened under the current legislation, the Copyright Designs and Patents Act 1988, by the inclusion of 'moral rights', entitling him to be identified as author and to object to derogatory treatment of his work. These rights developed first in Europe, arising from the idea that copyright was concerned with the protection of a work as an expression of its author's personality; their introduction in the United Kingdom, where protection of the author's economic interests has tended hitherto to receive most emphasis, is part of a growing convergence of basic principles in European copyright laws (Sherman and Strowel 1994; Bently 1994; Cornish 1995). The Copyright Act 1956, which replaced the 1911

Act and was the predecessor of the 1988 Act, recognised the development of new media in which the fruits of intellectual endeavour might be presented to the public – cinema films, television and sound broadcasting, cable television – by granting copyrights to the organisations responsible for their making. In 1984 the Cable and Broadcasting Act gave copyright to the signals sent to satellites for broadcasting back to earth. The following year computer software was recognised legislatively as having copyright, under the Copyright (Computer Software) Amendment Act. Not all the other problems concerning copyright in computer software were thereby resolved. Some of them were tackled, however, in the Copyright Designs and Patents Act 1988, which has in turn been amended under an EC Directive of 1991. The most recent concerns have been the application of copyright to digital technology and the "Internet" or "information super-highway", issues which have yet to be resolved (Dixon and Self 1994; Cornish 1995).

The advances of technology which made these new media possible also made copying in them easier and cheaper. Photocopying and tape recording were prominent early examples. The latter was particularly important because the most significant role of the new media was in the entertainment industry. Tape recording machinery became readily available to those who were the consumers of what the entertainment industry had to offer, so that home taping became the first serious challenge for copyright owners in the sound recording industry (and to a lesser extent in the sound broadcasting industry). While a photocopier is not yet a normal item of domestic furniture, its business and educational use has so increased that it now poses serious problems for authors and publishers aiming to serve those markets. The emergence of video recorders and personal computers as standard consumer items has created fresh difficulties, with copying, particularly in the case of computer software, being a matter of pushing a few buttons, switches and keys, with the result emerging sometimes only in a few seconds. The same applies to "downloading" material from the "Internet" and databases. Nor does the threat come simply from "home" copying. There is a market for the products of the entertainment and information industries which, for one reason or another, does not buy the originals and which does not have access to the machinery which makes home copying possible. It is this market which is fed by the "pirates", the organisations which exploit copying technology to put copies on the market at prices below those charged by the originators.

## Defending Copyrights

The range of rights with which the copyright owner has been able to battle the increasing scope for unauthorised copying and profit-making therefrom has always been formidable and is increasing. Sections 16-21 of the 1988 Act list the acts for the doing of which in relation to a copyright work the consent of the copyright owner must be obtained. Broadly, and *mutatis mutandis*, most of these have been infringements of copyright since 1911, if not before. They include copying the work, which is defined as substantial reproduction; that

## 10 COPYRIGHT, COMPETITION AND INDUSTRIAL DESIGN

is to say, the copying need not be exact or of the whole work. The reproduction may be in any material form: it need not be in the same form as the original work. Issuing copies of the work to the public without authority, performing, showing or playing the work in public, broadcasting it, or making an adaptation of it, or doing any of the other restricted acts in relation to an adaptation; all of these may be prevented by the copyright owner. He is thus in an excellent position to control the reproduction and marketing of his work, whatever it may have been.

The position of the copyright owner is further strengthened by the international structure of copyright law (Feather 1988: 172-4; Ricketson 1987; Stewart 1989). The Berne Convention of 1886 was an agreement between various States whereby the personal connection of an author with, or the first publication of a work in, a member State enabled the work to receive copyright in the other member States in accordance with the laws of those States. Accordingly, a copyright owner in a Berne country can have copyright around the world; indeed with the accession of the United States to the Berne Convention in 1989, there are few countries of significance outwith the system. While the laws of the member States are not identical, the Convention lays down the core of principle to which national laws must approximate. As a result, the copyright owner can assume a broad similarity of protection wherever it is required. Similar principles underlie the Universal Copyright Convention of 1952, which was designed to accommodate the United States within an international copyright framework. This international framework enables economically dominant countries to dictate the development of the law: a recent, although admittedly special, example in a field closely related to copyright is the Semiconductor Chip Protection Act 1984 in the United States, under which foreign producers of semiconductor chips were not entitled to protection in the States unless their domestic law accorded equivalent rights to American producers. The real target of the Act was Japan, whose producers were enjoying considerable success in the US market, rather than the developing countries. But since that market is the world's largest for semiconductor chips, the Act led to legislative action around the world; an excellent example of how intellectual property can be deployed in a protectionist way not only by proprietors but also by states.

The international framework for copyright and other forms of intellectual property was reinforced in 1993-94 by the conclusion of the Agreement on Trade-Related Aspects of Intellectual Property Rights (TRIPS) within the Uruguay Round of the GATT Agreement. This sets the minimum standard of copyright law for all countries wishing to participate in the world trading system, with sanctions to be imposed on those failing to comply. Computer programs and databases are to be protected; rental rights in computer programs and films are to be recognised; protection must last for the Berne minimum of the author's lifetime plus fifty years or, in the case of media copyrights, fifty years from creation or publication; and performers, phonogram producers and broadcasters are to receive rights. Much of this has already been achieved in Western legal systems, and it can again be argued that the effect of TRIPS, like other international agreements, is largely to benefit their developed economies rather than to facilitate the flow of ideas and information to the less advantaged parts of the world. Thus for example the United States' conflict with China early in 1995 over pirating of its intellectual property was in essence a protection of the American economy from the competition provided by the Chinese, who would only be allowed to enter the market upon the former's terms (see further below, 14-15).

This international framework of copyright law can be linked in the United Kingdom to what sections 22-27 of the 1988 Act describe as the 'secondary' infringements. As a result of the structure of international conventions, copies made overseas can be infringing copies in the United Kingdom. The main instance of secondary infringement is the importation otherwise than for the importer's private or domestic use of infringing copies; the others concern forms of dealing in infringing copies in the course of a business. The importer or dealer must know, or have cause to know, that the copies in question are infringing copies, whereas with ordinary infringement it is not necessary to show that the infringer knew that the work which he was copying was in copyright; this hurdle is easily overcome by the copyright owner sending the importer or dealer a warning letter to fix him with the necessary knowledge.

Yet these far-reaching rights and powers have apparently not proved sufficient for the protection of the copyright owner against infringement. The last twenty-five years have seen a steady development of his armoury against the infringer. The most striking instance of this came from the courts in England, with the 'Anton Piller' order, so named from the case in which it was first granted (Staines 1983). This enabled the copyright owner to gain access to information in the possession of possible infringers, without them receiving the opportunity to defend the owner's initial claim in court or to be warned in advance of the arrival of the court officers to execute the order. In 1981 the House of Lords held that a defendant might refuse to comply with an 'Anton Piller' order on the grounds that otherwise he might incriminate himself – a privilege well-established in the general law of evidence (*Rank Film Distributors Ltd v Video Information Centre* [1982] AC 380). The threat which this posed to the viability of the whole 'Anton Piller' scheme was such that Parliament acted to overrule the case in section 72 of the Supreme Court Act 1981. Procedures similar to the 'Anton Piller' order have been developed elsewhere in the legal world, including Scotland. There too it was held that the privilege against self-incrimination might be taken against a court order of this type (*BPI v Cohen* 1983 SLT 137) and again, albeit with slightly less haste, the legislature stepped in to restore the position of the copyright owner (Law Reform (Miscellaneous Provisions) (Scotland) Act 1985, s.15).

The relevance of the privilege against self-incrimination in this context is that in certain circumstances copyright infringement and dealing in infringing copies can be criminal offences. Under section 21 of the 1956 Act, however, the penalties were mild: a £50 fine for a first offence, and an option between a fine of the same amount and imprisonment for not more than two months for subsequent convictions. But a series of Acts in the early 1980s strengthened section 21 considerably, and sections 107-115 of the 1988 Act make very substantial provision on the matter. The maximum term of imprisonment under the Act is now two years and fines can reach the statutory maximum (currently £5,000 in summary proceedings in Scotland). Search warrants authorising the police to enter and search premises, using such reasonable force as is necessary, may be granted in respect of suspected offences while the copyright owner may also obtain the assistance of the Commissioners of Customs and Excise in prohibiting the importation of infringing copies. The removal from copyright offences of the privilege against self-incrimination, which can be linked to such fundamental ideas of civil liberties as the right to silence and the presumption of innocence, is thus no light matter.

## Policy Reasons Favouring Copyright

It is not difficult to see why courts and legislatures have moved to strengthen the position of the copyright owner in these ways. The traditional view, that copyright by guaranteeing a reward commensurate with the public demand for a work is an incentive to production which is itself in the public interest, remains a powerful one (Davies 1994). There is of course distaste for the unauthorised copyist who makes use of matter created by another; a distaste which is increased where the copying is on a large scale and for profits which might otherwise go to the originator of the work in question. It is this which justifies the description of the large-scale infringer as a child-stealer or pirate, robbing others of the offspring of their intellects and growing wealthy on the proceeds. But there is more to it than just a sense that infringement is morally reprehensible. The most recent study (Price 1993) suggests that in 1990 "industries with primary direct dependence on copyright accounted for 3.6% of GDP" (Gross Domestic Product, a measure of the total annual output of goods and services produced by United Kingdom residents). The industries contributed over £17billion to the GVA (Gross Value Added, a measure of the contribution of a particular industry to a good or service) in 1990 and employed over 800,000 people. They included printing and publishing, computer services, sound recording, film and broadcasting, retailing of copyright material, and performance rights. If industries "substantially dependent" on copyright are taken into account, the percentage contribution to GDP grows to 5.4% with employment of 1.3million. Although the exact contribution of copyright to these industries cannot be measured merely from these figures, since we do not know what the contribution of these industries to the GDP would be in the absence of the right, it is clear that this branch of the law relates to what in aggregate is a significant part of the British economy. Protection of the position of these industries through copyright is therefore most likely to be in the general economic interest. This conclusion was reinforced by Jennifer Skilbeck's (1988) study showing that copyright-using industries also made a major and rapidly growing contribution to British exports.

Similar observations have been made in other parts of the world (Price 1993: 12-13; Davies 1994: 69-70), while in a Green Paper the European Commission (1988) suggested that the service sector of industry, which includes the provision of information and entertainment, offers the best hope of economic expansion in the industrialised countries, where there has been a shift away from manufacturing and considerable new investment in the service industries. However, "those industries are also particularly vulnerable to damage through misappropriation ... [and] are those which are particularly exposed to losses through copying" (para 1.2.3). The Green Paper contains some estimates of the scale of these losses in respect of phonograms, films, video recordings and computer programs (para 2.2). While the figures are necessarily estimates, coming moreover from the organisations representing the interests of the copyright owners which have no reason to minimise the problem, it is clear that the losses have been substantial. The Paper also highlights the continuing advance of copying technology, making the process of piracy easier and better in its results, as a factor causing concern for the future health of the entertainment and information industries.

The 1988 Green Paper has been the basis for a subsequent programme of Directives, pursuing the harmonisation and strengthening of copyright within the Union. The first and most controversial concerned copyright in computer programs, and there are a number of others at varying stages of the European legislative process: one on rental rights, one extending the term of protection to author's lifetime plus seventy years, one to protect databases, and another on broadcasting rights. Although initially harmonisation in these areas was seen as necessary to prevent differences in national laws acting as barriers to the free movement of goods and services (see further below, 20), the overall effect of these Directives is clearly to enhance the protection offered by copyright (Cohen Jehoram 1994a). This is most obvious in relation to the Term Directive, which extends the period of protection for authors to lifetime plus seventy years; but the Rental Right and Database Directives also contain significant additional rights for copyright owners (MacQueen 1994).

United Kingdom legislation and European Directives have been responses to the lobbying of copyright owners and the organisations which they have formed to protect and promote their interests. Examples of such organisations from Britain include British Phonographic Industries, representing the sound recording industry, the Federation against Copyright Theft (FACT) and the Federation against Software Theft (FAST), representing the film and computer industries, and the British Copyright Council and the Copyright Licensing Agency, both representing publishers. The pressure which they have exerted has by no means been invariably successful – witness the failure in the United Kingdom to achieve the levy on blank tapes which at one time appeared likely to form part of the 1988 Copyright Act – but on the whole their record in having lobbies translate into legislation has been impressive. It is noteworthy that the blank tape levy is now on the European agenda.

All this is a far cry from the days of Lord Kames and Adam Smith, suspicious that undue extension of copyright beyond the period of protection laid down in the Act of 1709 would lead to monopolies, with all their bad consequences. In general, the eighteenth-century view of copyright has tended to be left behind in the development of the modern law. Its scope has been expanded legislatively and the courts have followed suit in their interpretation of its provisions. The result is that copyright is an extremely powerful form of legal protection and there is every prospect of its being developed further in the future.

## Challenging Copyright

But questioning voices echoing those of Kames and Smith have been heard to counter the widespread acceptance that the existence and further expansion of copyright is in the general economic interest. An extreme view would be that copyright is solely the creation of statute, amounting to no more than intervention by the State in order to restrict freedom and skew the operations of the market for the benefit of producers rather than consumers. It has been forcefully argued that if a justification for copyright is that by offering the prospect of reward it encourages authors to create works for the benefit of the public, then there is little evidence of its actually having that effect. The main beneficiaries of copyright are the entrepreneurial publishers rather than the creators of works; and it can be said as a matter of fact that most publishers seek to recoup their investment (including the payment of the creators) within a few years of publication, and do not base the decision on whether or not to publish a given work on the returns which may be earned during the full copyright term. There is accordingly little justification for the current length of the period of protection, and none for extending it, as proposed in the European Union. But the period of protection may not be merely about the economic return and the provision of an incentive to produce; it may also reflect a cultural policy which approves of creativity generally, without seeking to distinguish between which forms of creativity deserve protection (hence the low threshold of 'originality'). There is of course concern that without copyright the market would be open to "free-riders" who would not incur costs of production above the cost of copying, and would therefore be able to undercut the market price of the author and first publisher to the point where, since costs of production could never be recouped, it would become uneconomic for them to produce at all. It can then be argued that copyright protection, creating a sort of artificial "scarcity" of information, ideas and entertainment material, is a necessary condition before there can be any sort of market in this area. This has been met, however, by the comment that in the absence of competition the creator and the publisher can set a monopoly price for the work, limit the number of copies which are available to the public and curtail consumer demand. The evidence that this is the case lies in the widespread phenomenon of piracy, which would not exist unless there was a demand which the first producers were failing to meet. The competitive advantage of being first to market should not be underestimated; and the way to minimise the damage which may be caused by the "free-rider" is to set the initial price low enough so that there is little margin left to be exploited by the pirate. The problem then is that the return to the author and publisher may be very low. (On all this see Plant 1934, 1953; Hurt and Schuchman 1965; Breyer 1970, 1972; Tyerman 1971; Landes and Posner 1989; Puri 1990; Palmer 1990.) A further problem with this argument, however, is that the ease and speed of copying from and through modern technology means that "leadtime" arising through being first to market can be virtually non-existent, particularly in a global economy where it may be extremely difficult for the producer to enter all possible markets for the product at the same time. Copyright may thus help provide leadtime which the market cannot (Reichman 1994).

There is in any event a powerful argument that copyright is not a form of monopoly in that it only prevents competition by means of substantial reproduction where there is the causal link of copying. It does not prevent the independent production of the same work, unlike patent law, and it does not prevent an independent expression of the information and ideas embodied in the first work even where that work is the source for the second. Because close substitutes can therefore be provided and found, in most markets affected by copyright there is competition and consumer choice: for example, textbooks and sound recordings. Thus the market is usually contestable and the scope for monopoly pricing is lessened. Copyright also encourages competition by innovation as opposed to imitation. But this argument may break down where the work in question is expressed in what is really or practically the only possible form of expression: for example, the listings in a telephone or other directory of names and addresses, which require an alphabetical ordering to be of any utility. In such circumstances, and assuming the work to have copyright in the first place, the first to create may indeed have a monopoly (see further below, 22-3).

## Limiting Copyright: Fairness and the Public Interest

Gillian Davies (1994) has argued forcefully that the basis of copyright protection has always been in a concept of the public interest which seeks a balance between the interests of the individual and those of society at large. The grant of copyright to individual producers is because their work is seen to be for the social good, but at the same time the protection has been restricted or removed where some other, more weighty public interest has been involved. Thus in the United Kingdom copyright legislation since 1911 has recognised that the public interest may favour unlicensed copying in certain circumstances – for example, in reporting news, in an educational context, or for purposes of instruction, research and criticism. The 1988 Act also says that nothing in the Act "affects any rule of law preventing or restricting the enforcement of copyright, on grounds of public interest or otherwise" (s.171(3)); but it is significant that to some extent this express reference to the public interest had to be forced upon the Government during the Act's passage through Parliament (*Parliamentary Debates*, 8 December 1987, HL, cols 75-8; 23 February 1988, HL, cols 1162-4; 29 March 1988, HL, cols 630-4). The provision refers to a vague notion in the law that certain types of work may be denied the protection of copyright. In modern times the principal authority was *Glyn v Weston Feature Films* [1916] Ch 261, where the plaintiff's romantic novel *Three Weeks*, which in the words of the judge described "a sensual adulterous intrigue", was "grossly immoral" and "advocated free love and justified adultery where the marriage tie had become merely irksome", was denied copyright and protection against a satirical spoof in film called *Pimple's Three Weeks (without the option)* (which the judge thought "vulgar to an almost inconceivable degree"). There was doubt at one time whether *Glyn* was still of authority, given changes in mores since 1916, but the case was cited with approval by the House of Lords in 1988. This should not be seen as a signal for repressive backlash: the citation comes in the speeches in the great *Spycatcher* case, where Lords Keith, Griffiths and Jauncey all suggested that Mr Peter Wright would be unable to claim copyright in his memoirs, which broke the confidence of his former employers MI5. The case actually extends *Glyn*, inasmuch as Wright's immorality was not of a sexual nature; on the other hand, the speeches also suggest, not that the work had no copyright, but rather that Wright was not entitled to it. The copyright belonged to the Crown (*Attorney General v Guardian Newspapers Ltd (No 2)* [1988] 3 All ER 545). It may be suggested that the use of 'public interest' as a basis for challenging copyright is unlikely to become widespread as a result of its appearance in the 1988 Act.

British copyright legislation has also imposed controls from time to time upon how copyrights may be exploited through licensing. The 1911 Act, "show[ing] a percipient appreciation of the considerable market power that the most popular works would enjoy as the new medium [*of sound recordings*] developed its potential" (Cornish 1980: 397-8), introduced what was known as the 'mechanical right' to make recordings of musical works. When a record of a musical work had been made with the licence of the copyright owner, it was legitimate for others to make a recording of the same work, provided that a royalty at the rate of 6.25% of the ordinary retail price was paid to the copyright owner. This figure provided a ceiling on royalty rates for recording music and ensured accessibility of the material. Although the distinguished economist Sir Arnold Plant (1934: 194-5; 1953: 15-18) argued for the extension of this idea to other forms of publishing copyright works, the 'mechanical right' was abolished under the 1988 Act as an undue restraint on market freedoms.

The 1956 Act set up the Performing Right Tribunal to deal with complaints from would-be users for purposes of public performance of literary, dramatic or musical works, sound recordings and television broadcasts. Complaints had to concern the conduct of a collective licensing organisation in either refusing a licence or offering one on unreasonable terms. The Tribunal was to examine what was reasonable in the circumstances and had extensive power to make appropriate orders to give effect to its conclusions. Collective licensing is of crucial importance to the copyright in works intended for performance, in particular music and drama. Normally individual authors and composers transfer their rights to a collective organisation, such as the Performing Right Society, which can then be approached to license performances of the works. The composer receives a royalty from the Society in respect of licences granted. The Tribunal dealt with thirty-eight cases in the thirty-two years of its existence, a not overly-impressive workload. Its capacity to act as a watchdog over the exploitation of copyright through licensing was limited by its inability to take the initiative on unfair practices and by the limited scope of its jurisdiction. Under the 1988 Act the functions of the Tribunal now include licences for acts other than performance and it has been renamed the Copyright Tribunal. It can be said that in its new guise it has so far been much more active, albeit usually as not much more than a fixer of royalty rates (see MacQueen and Peacock 1995).

## Competition Law and the Copyright Monopoly: The UK Response

The public interest in restricting the scope of copyright has not been left simply to be defined by copyright legislation itself. On several occasions the exercise of copyrights has been referred to the Monopolies Commission, for consideration as to whether the conduct of the owners in question was against the public interest under competition law.[2] One case, to which we will return in detail later (see below, 40-2), concerned the refusal of the Ford Motor Company to license others to produce replacement body panels for its cars. Ford claimed copyright in the design of the panels and threatened infringement actions against any organisation seeking to make them. In 1985 the Commission found that copyright was being exercised in a way contrary to the public interest (Monopolies Commission 1985a). A different conclusion was reached by a majority in the *Radio Times/TV Times* report of the same year, which concerned the copyright in the magazines *Radio Times* and *TV Times* – or rather, the copyright in the programme schedules of the BBC and ITP, publishers of the magazines, which had been established in *BBC v Wireless League Gazette Publishing Co* [1926] Ch 433 and reaffirmed in *Independent Television Publications Ltd v Time Out* [1984] FSR 64. Both broadcasting organisations limited the extent to which others such as newspapers might publish advance programme information, so that the only publications which could publish the schedules for an entire week in advance were their own (Monopolies Commission 1985b; see further below, 18, 21-3). In 1988 the Commission reported on collective administration of the copyright in sound recordings and the licensing of broadcasts and public performances thereof. The Commission found that Phonographic Performance Limited, a collective licensing organisation for the owners of copyright in sound recordings, enjoyed a monopoly in the field which might be exercised in a way contrary to the public interest, although it accepted that "collective licensing bodies are the best available mechanism for licensing sound recordings provided they can be restrained from using their monopoly unfairly". It went on to make various recommendations to ensure this (Monopolies Commission 1988).[3] Finally, in 1994 the Monopolies Commission rejected the contention of the National Heritage Select Committee that the ability of recording companies to use their copyrights to restrain parallel importing of cheaper records and compact discs from the United States was against the public interest (Monopolies Commission 1994).

We need not concern ourselves further here with the details and reasoning of the reports: the immediate interest lies in the revival of the old arguments about the capacity of copyright to create a monopoly and about the public interest in these matters. And it is clear from the course of the debate that the unfettered exercise of copyright is not invariably the answer to this particular question. Indeed, on the matter of broadcasting schedules the British Government showed that it dissented from the Monopolies Commission conclusion that the exercise of the copyright was not against the public interest by legislating in the Broadcasting Act 1990 to require the BBC and ITP to grant licences and provide advance information about programme schedules. The result has been the appearance on the market of a number of competitors for *Radio Times* and *TV Times*, although the terms and conditions of the licences were only settled finally after the intervention of the Copyright Tribunal (*News Group Newspapers Ltd v Independent Television Publications Ltd* [1993] RPC 173). In addition the 1988 Act confers power upon the Secretary of State for Trade to make orders in respect of licences found by the Monopolies Commission to include conditions adverse to the public interest, or where the refusal to grant licences is found to operate against the public interest by the Commission. This fills a lacuna in the law which had become apparent as a result of the Ford and *Radio Times/TV Times* reports.

Perhaps influenced by some of this activity, there were a number of judicial pronouncements in the 1980s criticising the monopolies conferred by copyright and other forms of intellectual property. The most striking instance was in the case of *British Leyland v Armstrong Patents* [1986] AC 577, in which Lords Templeman and Bridge in particular were strongly against what they saw as the monopoly exploitation of copyright (see further below, 45-7). Similar remarks by these two judges can be found in other cases of the period. In *Re Coca Cola Co* [1986] 2 All ER 274 at 276, Lord Templeman said of an application to register the Coke bottle as a trade mark that it was "another attempt to expand on the boundaries of intellectual property and to convert a protective law into a source of monopoly ... This raises the spectre of a total and perpetual monopoly in containers and articles". In *CBS Songs Ltd v Amstrad Consumer Electronics plc* [1988] AC 1013 at 1060, a case which concerned home taping of sound recordings, he observed that "some home copiers may consider that the entertainment and recording industry already exhibit all the characteristics of undesirable monopoly, lavish expenses, extravagant earnings and exorbitant profits". For his part, Lord Bridge commented in *Green v Broadcasting Corporation of New Zealand* [1989] 2 All ER 1056 at 1058 that "the protection which copyright gives creates a monopoly and there must be certainty in the subject matter of such monopoly in order to avoid injustice to the rest of the world".

## Competition Law and the Copyright Monopoly: The EU Response

The establishment of the EEC by the Treaty of Rome in 1957 was intended to establish a 'common market' amongst the Member States by, *inter alia*, eliminating customs duties and 'quantitative restrictions' on the import and export of goods between them, and also by removing all obstacles to the free movement of goods, services and capital within the market. A system to ensure undistorted competition throughout the market was to be established and the laws of Member States were to be approximated to the extent necessary to permit the proper functioning of a common market. Intellectual property rights and their exercise presented a considerable obstacle to this goal, given that the rights depended primarily upon national legislation which was very varied in content and gave right-owners the ability to control distribution and supply of products infringing those rights within the common market on a national basis. Yet it was also clear that the rights were of such economic significance in the Member States that a process of reconciliation between them and the objective of a common market, rather than their complete removal, would be necessary.

The process of reconciliation has gone on at various levels within the Community since its foundation (see MacQueen *et al* 1993: paras 1611-1664). The EC Treaty itself states in Article 36 that 'industrial and commercial property rights' may justify restrictions on the free movement of goods, which is consistent with the general recognition in Article 222 that property rights are unaffected by the Treaty. In the long term, perhaps the most significant steps have been those taken towards a system of Community intellectual property. Thus in 1975 the Member States signed the Community Patent Convention by which it will be possible to obtain a single patent which can be granted, transferred, revoked or allowed to lapse only in respect of the whole Community. As a result of constitutional difficulties in some Member States, this has still to come into force despite much effort. There has been more success in establishing a Community Trade Mark. A Community Trade Mark Office was opened in Alicante in 1994, and the first Community Trade Marks may be granted in 1996. Inasmuch as the variations of national intellectual property laws have shown themselves capable of forming barriers to the free movement of goods and services, there has also been a programme of harmonisation through Directives. The principal examples are in the fields of trade marks and (as already noted) copyright; but the harmonisation programme has also had some setbacks, most notably in February 1995, when the European Parliament rejected a draft Directive proposing extended patent protection for biotechnology products.

It has been remarked, however, that the work of the European Court of Justice, in considering the impact of intellectual property rights upon the common market, has already produced the results which such Community systems of intellectual property would wish to achieve – that is, permitting the existence and exercise of intellectual property rights within the common market only insofar as consistent with the principles of the EC Treaty. In the context of the free movement of goods, Article 36 has been read as meaning that intellectual property rights may be exercised in some circumstances, but not all. Where the owner has 'exhausted his rights', he may exercise them no further. The Court has developed the concept of the 'specific subject matter' of intellectual property rights, stating that the owner may exercise his rights only so far as necessary to protect their specific subject matter. Beyond that his rights are said to be 'exhausted' and their exercise, even if permitted by national law, will fall foul of Community law on the free movement of goods. The specific subject matter of most intellectual property rights has been held to be the right to put the product associated with the right – the patented invention, the trade-marked goods, the copyright book – into circulation for the first time, or to license others to do so. Accordingly if A in Member State X licenses B to market the product in Member State Y, he cannot object when C imports into State X products bought from B in State Y. A has exhausted his rights in connection with those goods because he licensed their first marketing in State Y.

A similar distinction, between the existence of an intellectual property right and its exercise, has been drawn in applying the competition rules of Articles 85 and 86, which prohibit, respectively, cartels and agreements distorting competition and abuses of dominant positions within the common market. While the existence of intellectual property rights does not automatically infringe these Articles, their exercise – principally through licensing schemes or refusals to license – may do so. This distinction of existence from exercise has been rightly criticised – "a right cannot consist of more than the various ways in which it can be exercised" (Korah 1972: 636) – but by generalising the approach originally adopted in the interpretation of Article 36 across the whole field, the Court and Commission have attempted to balance the rights conferred by national laws with the common market. Ground rules have also been established by the European Commission to exempt patent and know-how licensing and franchising schemes from the operation of Community competition law, provided that certain types of clause are not used.

The limits of these principles in achieving the goal of a single market have, however, become more apparent since the mid-1980s, particularly where differences between national laws are concerned. The Court has become more reluctant to over-ride national laws, unless they actively discriminate against the nationals of other Member States (see e.g. *Phil Collins v Imtrat* [1993] 3 CMLR 773). This has compelled the Commission and Council to pursue more energetically the deepening of the single market through the harmonisation programme and the creation of Community intellectual property rights. Thus in *EMI Electrola GmbH v Patricia Im- und Export* [1989] ECR 79 sound recordings in which the copyright had expired in Germany were lawfully manufactured there, but it was held that the copyright still surviving in Denmark could be used to prevent their import into that country. It was this case which led to the Term Directive, just as the successful invocation of Danish rental right in *Warner Bros Inc v Christiansen* [1988] ECR 2605 to prevent the import of videos from the United Kingdom, which then gave no equivalent protection, lies behind the Rental Right Directive.

Two of the cases examined by the Monopolies Commission in the United Kingdom have been considered further under the competition laws of the European Union, those concerning Ford spare parts and TV programme schedules. The former was settled in 1990, when Ford gave undertakings to the Commission that it would not insist upon absolute exclusivity in the exercise of its rights and that it would grant licences on reasonable terms ([1990] 5 EIPR D-101). The approach clearly indicated that full-blown exercise of intellectual property rights to exclude others from the supply of replacement parts for motor vehicles was anti-competitive. (For further discussion, see below, 42-5.) On the question of copyright in broadcasting schedules, the United Kingdom legislation of 1990 came too late to prevent the Commission from investigating the matter; in any event there was a similar problem in Ireland, raised with the Commission by a company called Magill, for whom the case is generally named. In 1988 the Commission ruled that the refusal of broadcasting organisations to license other publishers to produce advance weekly listings was an abuse of a dominant position contrary to Article 86 of the EC Treaty. This ruling was upheld by the Court of First Instance in 1991 (Case T-69/89, *Raidio Telefis Eireann v Commission* [1991] ECR II-485). In June 1994 Advocate General Gulmann gave his opinion in the case (Joined Cases C-241/91 P and C-242/91 P), and argued that the decisions below should be over-turned. This advice was not accepted by the Court of Justice, which affirmed the decisions of the Commission and the Court of First Instance on 6 April 1995.

The central theme of the Advocate General's opinion was that the exclusive right conferred by copyright should not by itself be regarded as giving a 'dominant position' for the purposes of competition law. It acknowledged that there is no meaningful distinction to be drawn between the existence and exercise of rights. Part of the essence of copyright is to be able to exclude others from the owner's marketplace. A decision that the copyright owner's refusal to grant licences to others is an abuse meant imposing compulsory licences and encroaching upon the specific subject matter of copyright. Copyright is not simply about earning a reward for the owner, as might still happen with a compulsory licence; it embraces the author's moral right to control reproduction. Interference can only be justified where there are special circumstances, which the Advocate General did not find to exist in the case before him. Ultimately he did not define what 'special circumstances' might be, preferring to leave that to case-by-case development.

The Court of Justice accepted that mere ownership of an intellectual property right did not by itself confer a dominant position under Article 86. However in this case the broadcasting companies were the only source of information about programme schedules and as a result were in a position to prevent effective competition on the market for weekly television magazines. The exercise of an intellectual property right might, in exceptional circumstances, involve abusive conduct. Here the refusal to license had prevented the appearance of a new product for which there was consumer demand; the refusal could not be justified in terms of broadcasting or publishing activities; and the right-holders had excluded all competition by denying access to the raw material indispensable for the compilation of a programme guide. Finally, the Court observed that since the European Community is not a party to the Berne Convention, Article 9 thereof, which allows derogations from a copyright owner's exclusive right of reproduction only if it does not conflict with normal exploitation of the work or prejudice the author's legitimate interests, could not be relied upon to prevent the ordering of compulsory licences.

The decision of the Court is extremely important and affirms the applicability of competition law to the exercise of copyright. It is bound to be controversial, as its effect is to impose a compulsory licence upon a copyright owner (see Haines 1994; Miller 1994). But it should obviously not be seen as meaning that competition law applies every time a copyright owner decides not to grant licences. Like the Advocate General, the Court states that there must be 'exceptional circumstances' before competition law steps in. In this particular case, the exceptional circumstance was the inability of potential competitors to express the information concerned in any other way, even if they had been able to gain access to it. The copyright conferred a monopoly in information which for practical purposes could only be expressed in one way. At all stages in the case there was hostility to the notion that there could be copyright in lists of programmes. Even the Advocate General indicated as much:

> It seems to me that the Commission's decision and the judgments of the Court of First Instance produce a reasonable result in practice. There are strong reasons to suggest that it should not be possible for television broadcasting organisations to prevent, by means of their copyright in programme listings, publication of comprehensive weekly television guides. I do not consider that the copyright interests thus protected can be regarded as substantial, and the Irish and United Kingdom consumers have a clear interest in being given access to a product which is common in the other Member States and which offers a number of advantages by comparison with existing products.

There are also echoes here of the decision of the US Supreme Court in *Feist Publications Inc v Rural Telephone Service Company Inc* 113 L Ed 2d 358 (1991) that a list of telephone subscribers, arranged in alphabetical order, lacked even the minimum degree of creative spark to enjoy copyright, despite the considerable effort and expense involved in its production.

The key to understanding the likely future interplay between copyright and competition laws in the European Union may perhaps still be found in a passage in the Green Paper (1988: paras 1.3.5-6):

> [C]opyright is an exclusive right granted by legislation to an individual. *One of its effects is inevitably to limit to a certain extent the normal freedom of third parties to compete by marketing similar products. In the more traditional domains of copyright applying to literary, musical and dramatic works, this has not posed a significant problem since independent works of the same genre can in law and practice still compete with each other quite fairly* [emphasis supplied]. In areas which have developed more recently, however, the restrictive effects of copyright protection on legitimate competition have on occasion risked becoming excessive, for example, in respect of purely functional industrial designs and computer programs. In such contexts, copyright protection without suitable limits can in practice amount to a genuine monopoly, unduly broad in scope and lengthy in duration. It follows that, in developing Community measures on copyright, *due regard must be paid not only to the interests of the right holder but also to the interests of third parties and the public at large* [emphasis supplied], since, particularly with regard to products of an industrial character, works are placed on the market by a decision of the right holder himself.

The italicised passages demonstrate a recognition that it is only where there are restricted modes of expressing a work, as was the case with programme schedules, that copyright should be subject to other controls.

It is clear, therefore, that under Community law the rights conferred by copyright may still be over-ridden where there are countervailing aspects of what is taken to be the public interest – whether that is overcoming barriers to the creation of a single European market or to free competition on that market. There is a perception that not all the rights which a particular copyright law may confer are necessarily in the public interest. The difficulty lies in finding a way of testing the public interest in these matters the validity of which can be generally agreed.

## Conclusions

It is time now to draw the strands of this rather discursive chapter together and to point the way forward to the rest of the book. It has attempted to demonstrate that copyright is a powerful economic tool capable of yielding considerable rewards to those entitled to the benefit of its protection. It is of major importance to creative individuals in the fields of literature, drama, art and music but it is quite mistaken to think of copyright as being of significance only or mainly in these areas or for creative persons working in them. First it is not only about aesthetic creations; copyright protection is extended to all products of the appropriate form which are the result of skill and labour. Secondly copyright is of vital importance to the producers of the media in which so many works are communicated to their public – the publishers and the recording, film and broadcasting companies. For such organisations, copyright is not merely a method of obtaining a return for creative endeavour. The tool of copyright is also capable of use as a weapon in the market place, either to prevent, control, or eliminate competition. As such it is capable of abuse, as was recognised in the eighteenth century. The question of how to respond to this fact comes down to one of the public interest. The justification for copyright has always been the public interest and that remains the case. But it is easy to turn copyright into some sort of unchallengeable fundamental right the exercise of which is always justified. In at least some circumstances there may be countervailing considerations of the public interest, requiring a balancing act to be carried out before conclusions are reached. The problem which the preceding pages have shown as emerging increasingly clearly is whether this balancing act is to be achieved through the application of competition law principles, or by reflection on the substance of the law of copyright itself.

The remainder of this book examines a particular area of commerce in which these questions have been and continue to be fundamental issues. The area in question is that of industrial design – that is to say, the design of goods intended for industrial or mass production. Copyright was largely excluded here until this century when, thanks largely to the philosophy, "what is worth copying is worth protecting", it was brought under the protective umbrella. Now in the

United Kingdom it has been excluded again and a new form of protection, akin to copyright, has been introduced. Meantime in the European Union the whole question is being addressed anew. The aim of this book is to explore the issues raised by the protection of industrial designs, considering the matter primarily from the standpoint of British developments but concluding by examining what if any form European protection may take. The next two chapters consider the historical background to the current law in the United Kingdom, firstly because this is indispensable to an understanding of the present rules discussed in Chapter Four, but secondly because the debate is revealing and instructive as to the proper scope of copyright as a means of protection and its relation to the promotion of competition and other economic goals. This in turn throws light on the current state of play in Europe, discussed in Chapter Five, and suggests where the right balance may lie in determining whether protection should exist at all and, if so, whether by copyright or some other right giving the kind of protection which encourages and rewards competition by innovation rather than imitation.

# Chapter Two

## The Rise and Fall of Design Copyright

### Protection of Products Through Intellectual Property: The Traditional Techniques

The commercial problem which gives rise to the legal issues to be discussed in the rest of this book is in essence a simple one. How may a manufacturer stop imitation of his products by his competitors? Or, to put it another way, how may he inhibit others from competing by taking over his good ideas? The fact that he is the victim of the sincerest form of flattery will be of small comfort as his profit margins dip and those of his competitors increase. One answer, of course, might be that this kind of competition should not be stopped and that the manufacturer should instead seek to maintain or increase his profits by leading the market through a continual process of innovation. Another diametrically opposed view might argue that some basic right of property was being invaded: "of all things, the produce of a man's intellectual labour is most peculiarly distinguishable as his own", and so there should be legal redress when it is misappropriated by others. The law has for the last two or three hundred years taken a median position: certain types of product are protected against certain types of imitation and otherwise the market is free.

### (i) Patents

The classic form of legal protection for a product is the patent, now granted under the Patents Act 1977. For a period of twenty years from the date the successful application is filed, no-one may make, dispose of, offer to dispose of, use or import a patented product, or keep it whether for disposal or otherwise, unless he has the consent of the proprietor of the patent. A patent is granted for an invention which is new, involves an inventive step and is capable of industrial application. To be new an invention must not have been available to the public, whether by written or oral description, by use, or in any other way. To involve an inventive step, the invention must not be obvious to a person skilled in the art, having regard to such matter as has been previously available. These are stringently defined tests, stringently applied, and as a consequence it is both difficult and expensive to obtain a patent. Application must be made to the Patent Office in Newport or, under the European Patent Convention 1973 (see below 32), to the European Patent

Office in Munich, and must include "a specification containing a description of the invention, a claim or claims and any drawing referred to in the description or any claim"; this specification must "disclose the invention in a manner which is clear enough and complete enough for the invention to be performed by a person skilled in the art", while the claim must "define the matter for which the applicant seeks protection, be clear and concise [and] be supported by the description". (For the foregoing see Patents Act 1977 ss 1-3, 14 and 60.) Drafting the specification is generally a matter for the skilled assistance of a patent agent. Applications are published in the Patents Journal, to provide an opportunity for third parties to make observations on the patentability of the invention; this also enables competitors to take those good ideas which fail to achieve a patent, so that the making of an application becomes a delicate balancing act for the draftsman between disclosure sufficient to obtain the patent and concealment of details from business rivals. There is then an examination and search within the Patent Office, for which the applicant pays fees, to test the patentability of the invention. Since 1989 there has been a slow decline in the number of applications for patents at the Patent Office, and more recently at the European Patent Office also; but nevertheless there were 26,648 applications for a British patent in 1993, and 56,808 applications at the European Patent Office in the same period (Patent Office 1994: 11, 17).

Patents are difficult to obtain and difficult to defend. The classic modern illustration is *Windsurfing International Inc v Tabur Marine* [1985] RPC 59, where a 1968 patent for a windsurfer was successfully challenged because there had been prior publication in a periodical article two years before, and prior use by a 12-year old boy at Hayling Island ten years before the patent application was filed. But if the patent stands up, then it is a potentially wide-ranging weapon. The specification defines an area within which the patentee has a monopoly – he does not need to prove copying, simply making, using and so on. Moreover a purposive rather than a literal approach should be used in determining the scope of the specification. This was the ruling of the House of Lords in *Catnic Components Ltd v Hill & Smith* [1982] RPC 183, a case concerning load-bearing lintels. The plaintiff's specification indicated that the principal support should be supplied by a metal face 'extending vertically' from the base. The infringing lintel's support was at an angle of six degrees to the base but its load-bearing capacity was virtually identical. It was held that 'extending vertically' in the specification should not be confined to its literal meaning and infringement was established. This is also the approach to be taken under the European Patent Convention. The scope of patent protection is nonetheless limited by the requirements of patentability, and is unlikely to be of much assistance in respect of the majority of industrial products.

*(ii) Registered Trade Marks and Passing Off*

A second possibility is to use trade mark law. In essence this branch of intellectual property is about the protection of the badges of identity which traders and others use in connection with their goods and services to enable customers to distinguish their products from those of competitors. The scope for obtaining protection in this way through the common law passing off action is limited. The common law has been reluctant to recognise a product as comprising in its entirety a badge of identity, and has only been prepared to grant protection to the 'get up' and containers in which goods are marketed where these are recognised in the market as distinctive of a particular producer (MacQueen et al 1993: para 1380). The best-known example in recent times is the JIF lemon, which was held by the House of Lords to be a source of goodwill only for Reckitt & Colman, which could accordingly stop others marketing lemon juice in lemon-shaped containers (*Reckitt & Colman Products Ltd v Borden Inc* [1990] 1 All ER 873). It seems clear from several cases, however, the most recent involving the Cadbury's Chocolate Flake, that the shape of the product itself cannot be a badge of identity capable of protection by an action of passing off, although action may be possible in respect of some distinctive *part* of the product (MacQueen et al 1993: para 1380).

A British registered trade mark may now offer more scope for product protection than passing off, following the replacement of the Trade Marks Act 1938 by new legislation in 1994. A trade mark is now defined as any sign capable of being represented graphically and of distinguishing goods or services of one undertaking from those of other undertakings. In particular it may consist of the shape of goods or their packaging (1994 Act s 1(1)). The effect of this is to reverse the decision of the House of Lords in *Re Coca Cola Co* [1986] 2 AllER 274 under the 1938 Act, that a container such as a bottle could not be a registered trade mark, no matter either how distinctive of a particular manufacturer it might be, or that the law of passing off could have been invoked against any other user of a bottle in that shape. It is not anticipated, however, that this change in the law will mean that registration of product shapes and containers as trade marks will suddenly become the primary means of protection from imitation, any more than passing off has been hitherto. It will need to be clear at the time of application for registration that the shape in question already serves to link the goods with a particular producer in the public mind (Department of Trade and Industry 1990: para 2.13). There is also an important limitation in provisions that a sign is not to be registered as a trade mark if it consists exclusively of a shape which (a) results from the nature of the goods themselves; (b) is necessary to obtain a technical result; or (c) gives substantial value to the goods (Trade Marks Act 1994 s 3(2)). The aim here is obviously to prevent trade marks being used to prevent competition between products as such, although the scope of the exclusion and its relationship with the permission to register shapes and containers will probably only become clear in future case law (Department of Trade and Industry 1990: paras 2.16-2.21; Annand and Norman 1994: 84-87; Morcom 1994: 17-18; Institute of Trade Mark Agents 1994: 40-42).

Trade mark protection will have attractions for producers, however. Although registration must be in respect of particular goods or services, it lasts for ten years in the first instance and may subsequently be renewed for further periods of ten years at a time. There is no overall maximum period for a registration, which gives the proprietor the exclusive right to use the mark in the course of business in respect of the goods or services for which it is registered. Again, therefore, we have a monopoly right, comparable to that conferred by a patent. Use by another of a similar mark in respect of the same or similar goods or services, or of an identical mark in respect of similar goods or services, is also infringement of the proprietor's rights. Finally, use of an identical or similar mark in respect of goods or services which are not similar to those for which the trade mark is registered can also be infringement, if the trade mark has a reputation in the United Kingdom and the infringing use takes unfair advantage of, or is detrimental to, the distinctive character or repute of the trade mark. Registered trade mark protection is thus significantly wider in scope than was the case under the 1938 Act, and seems certain to be more attractive to business as a way of protecting product shape and get up. It is therefore possible that a trend of gradual decline in the number of applications for a British trade mark – from nearly 40,000 in 1989 to 34,764 in 1993 (Patent Office 1994: 20) – may be reversed, although that may be tempered by the availability from 1996 of the Community Trade Mark.

*(iii) Registered Designs*

A third way of protecting a product from appropriation by competitors is through the law of registered designs. Since 1839 it has been possible to register new and original designs which are to be applied industrially, a phrase which in modern law has meant "applied to more than fifty articles" (Copyright (Industrial Process and Excluded Articles) (No 2) Order, SI 1989/1070). At present registration is governed by the Registered Designs Act 1949. Prior to the Copyright Designs and Patents Act 1988, registration gave the owner of the design an exclusive right for up to three successive periods of five years from the date of registration to sell, make and import articles to which the design had been applied. The maximum period of protection was thus fifteen years, but the registration lapsed if it was not renewed at the end of any five-year period during which it had subsisted. The 1988 Act extended the period of protection to a maximum of twenty-five years in five five-year blocks for reasons to be discussed later (below, 67-8). But the possibility of obtaining this protection is restricted by certain requirements. To be registrable a design must be new and constitute features of shape, configuration, pattern or ornament applied to an article by any industrial process or means, being features which in the finished article appeal to and are judged solely by the eye (1949 Act s 1). This is the formula, more or less, which has applied since the Copyright in Designs Act 1842, which laid down that "new and original Design[s]... applicable for the Pattern or for the Shape or Configuration, or for the Ornament" of articles of manufacture might be registered. The formula was however developed in section 19 of the Patents and Designs Act 1919 which added the 'eye appeal' element and, even more importantly, excluded from the definition of design "any mode or principle of construction, or anthing which is in substance a mere mechanical device". In the 1949 Act this exclusion was rewritten so that features 'dictated solely by the function' to be performed by the article to be made in that shape or configuration are not designs for the purposes of the Act and are not registrable (1949 Act s 1).

Clearly, therefore, the registered designs legislation has at no time provided universal protection for industrial designs. Only those with eye appeal qualify and those which are functional are excluded. The distinction is one which cuts across much of the thinking which was going on about design and industrial products more or less contemporaneously with the coming into existence and development of the registration system. Nikolaus Pevsner (1975: 10) has observed that for much of the first century of the Industrial Revolution practically all industrial art was crude, vulgar, and overloaded with ornament". To some extent the stress of the 1842 Act on ornament, pattern, shape and configuration reflects this understanding of what was valuable in a design. A reaction set in, associated particularly with the names of William Morris and, in Scotland, Patrick Geddes, but developed by others in connection with the objects of mass production. Ornament, pattern and design not associated with function were to be abhorred and abjured and instead designers laid stress on the production of articles which were fit for their purpose. There was no conflict between art and utility; art lay in finding the form and materials appropriate to the function which the object was to perform. This was the credo of what has become known as the Modern Movement in design; it is ironic that at the very time when that credo was becoming orthodoxy, in the mid-twentieth century, registered designs legislation was beginning to state explicitly that only non-functional elements in a design were worthy of legal protection from imitation (McCarthy 1979; Huygen 1989).

The distinction between eye appeal and functionality has given rise to much complex and, on occasion, entertaining litigation. The most diverting cases are those about the designs for chocolate eggs and digital watch faces, in both of which the design in question was normally invisible. For the egg it was the design of inner and outer layers of the egg shell in contrasting tones of brown and white; for the watch it was the design of the liquid crystal display. It was argued that neither could 'appeal to the eye' as a result of their invisibility; it was held that the appeal to the eye had to be tested when the article was in use: with the egg, when it was being eaten, with the watch, when the display was lit up. (See *Ferrero and CSPA's Application* [1978] RPC 473; *KK Suwa Seikosha's Application* [1982] RPC 166.)

The leading cases are *Amp v Utilux* [1972] RPC 103, decided by the House of Lords, and *Interlego v Tyco Industries* [1989] AC 217, a ruling of the Privy Council on an appeal from Hong Kong. *Amp* established that the eye in question is that of the customer and that the appeal has to be through features calculated to attract custom. A design will be registrable when only some of its features are functional and others have eye appeal – the exclusion only operates when a design is wholly functional. A design will be taken as dictated solely by function even if the function could equally well be performed by an article in different shape. *Interlego* confirms all the points made in *Amp* and also elucidates a matter which that case had left unclear: what if there is a combination of eye appeal and functionality in a design? The conclusion of the court was that if all the features of a design were dictated by function and eye appeal was fortuitous, there could be no registration. In considering the fortuitousness or otherwise of the eye appeal, the intention of the designer was relevant, although not conclusive.

These general tests take on a greater content through consideration of the facts and decisions in the *Amp* and *Interlego* cases. The issue in *Amp* was the registrability of a design for electrical terminals used in washing machines. The terminals were shaped to enable them to hold electric leads. It was held that the design was dictated solely by function, even though it was possible to design the terminal in other ways in which it would still carry out its function. Accordingly it was not registrable. 'Dictated solely by function' did not mean that a design was unregistrable only if it was the one possible design, which had been the previous understanding of the law; it meant simply that the features of the design were there for the purpose of function alone. The features of the terminal were dictated solely by function in this sense. The test thus stated changed the law; as a result, many designs hitherto regarded as registrable ceased to be so. Further it was held that the design had no eye appeal: none of its features were there to attract the attention of the customer. True, the functionality of the design might appeal to customers considering the utility of the article but, in the view of the majority in the House of Lords, this was not how the test of eye appeal should be applied. Eye appeal meant that to be registrable the design should have features going beyond functionality and there to catch the eye of the customer.

Where *Amp* is primarily about functionality, *Interlego* concentrates on eye appeal. The case concerned the design of the well-known children's toy, the Lego building kit, which was held to be registrable. The problem which the court had to overcome to reach this conclusion was encapsulated thus by Lord Oliver ([1988] AC at 246):

> Inevitably a designer who sets out to make a model brick is going to end up producing a design, in essence brick shaped...There is clearly scope in the instant case for the argument that what gives the Lego brick its individuality and the originality without which it would fail for want of novelty as a registrable design is the presence of features which serve only the functional purpose of enabling to interlock effectively with the adjoining bricks above and below.

At the same time, however, the design clearly had eye appeal and Lord Oliver was able to hold that this was not fortuitous, the evidence led in the case showing that attraction of the customer had been a factor in the mind of the designer.

A further requirement before a design may be registered is that it should be 'new'. Morris and Quest (1987: 106) point out that the possible shades of meaning in that word:

> A design may be new in an absolute sense in that the shape or pattern has never been seen before; or it may be new in the limited sense that the shape or pattern already known, but only in a context different from that now contemplated. Also a design may be new in that it differs from known designs, but in a more restricted sense it may lack novelty because the difference is of an inconsequential nature.

What is clear, however, is that novelty is generally a more restrictive requirement than the traditional copyright test of originality, inasmuch as with the latter something may be original if it is produced by independent skill and labour, even though the end result is similar to an existing work. The requirement also means that the owner of the design cannot test the market for products made to the design in advance of registration, because publication in this way will deprive it of novelty.

Creating a registrable design is thus potentially problematic. In addition and quite apart from the legal tests of registrability, there are the procedural difficulties. Registration can only be carried out at the Designs Registry. It is part of the Patent Office, which was relocated in Newport, Gwent, in 1991, although it retains a London address.[1] Access therefore poses some difficulties, although applications may be made by post. Actual registration procedure is complex and time-consuming even when the registrability of the design is not a major issue; the interval between application and registration is likely to be at least six months. The assistance of a costly professional representative such as a patent agent will almost certainly be required. There are registration fees totalling typically a few hundred pounds to be paid at the Registry (see Registered Designs (Fees) Rules 1992, SI 1992/617), and the fees due to one's agent; and these costs are incurred without knowing whether or not the design is of any commercial value. While the time taken to register a design and the cost of doing so are much less than for an invention or a trade mark, nonetheless the process is fraught with potential difficulties and the benefit obtained is uncertain value for money, given the uncertainty in any event as to the design's commercial prospects before marketing. There are many times fewer applications for design registrations each year than for patents and trade marks (compare the figures for patents and trade marks quoted above, 26 and 28, with those for designs quoted below, 67); this suggests that registration is not seen as so important as the others, whatever may be the commercial significance of design itself.

This completes the survey of those areas of intellectual property law traditionally associated with industrial products. Copyright, as will be discussed at greater length shortly, was not operative in the area and the material reviewed thus far shows several broad distinctions that can be drawn between copyright and the other forms of intellectual property. First and foremost is the question of what may be called 'access'. It is much more difficult to get a patent, trade mark or a registered design than to acquire a copyright. The last-named comes into existence automatically with the creation of an original work; and originality, it will be recalled, is a test concerned primarily with an input of independent skill and labour by the author and not with the quality or merit of the work. For patents, trade marks and registered designs, registration is generally essential and conditions must be met: inventiveness; distinctiveness; eye appeal. Even the common law remedy of passing off requires the existence of goodwill and distinctiveness, so that even though registration is not required, there are still significant hurdles to be overcome. Registration being a complex process, professional assistance will almost certainly be required, which presents for some at least a further obstacle, which is not present in copyright law, to the realisation of protection.

The requirement of registration for patents, trade marks and registered designs also limits the scope of international protection by comparison with copyright. Where the latter arises automatically throughout the Berne Union, registration of a patent, trade mark or design in one country generally only establishes a short-lived right of priority in seeking registration in other countries, this arising under the Paris Convention for the Protection of Industrial Property established in 1883. There are international bureaux where an application to register a right in several different countries at once, created under various treaties and other agreements since then; but this is still costly and time-consuming, and will often still involve procedures in national registries. One of the most successful international offices is that for European patents in Munich, set up under the European Patent Convention 1973, where a bundle of national patents can be obtained in a single process; it provides a model for future developments. But meantime there can be no doubt of the advantages enjoyed by an international copyright owner by contrast with his counterpart in patents, trade marks and designs.

A second distinction is that patents, registered trade marks and registered designs are all 'monopoly' rights. The owner does not need to show that there has been copying before his action succeeds; all he needs to show is use of the same or a not substantially different product, mark or design by the other party. Even if the other party has achieved the result independently he is liable. This is quite different from copyright law, where the causal link of copying must be established before an action can succeed. Admittedly, in most copyright actions this link often cannot be categorically proved, and the proof rests on the circumstantial evidence of the claimant's work being first in point of time, the alleged infringer having had access to it before his own production, and the similarity of the two works. Nonetheless, this is more than will be needed with other forms of intellectual property. The nearest copyright law has got to the 'monopoly' right is to recognise the possibility that copying might result from the operation of the sub-conscious mind (*Francis Day & Hunter v Bron* [1963] Ch 587).

Finally, attention should be drawn to the much longer period of protection found in copyright law in comparison to the other forms of intellectual property. In particular, its typical formula of author's lifetime plus fifty (soon to be seventy) years contrasts sharply with the twenty years of the patent and the fifteen years of the registered design before the 1988 Act. Only trade marks can be protected for a period longer than copyright.

## The Development of Design Copyright

Design copyright did not arise as a possibility until the Copyright Act 1911. This was the first legislation to confer general copyright protection upon all original artistic works. Copyright had previously subsisted in sculptures (under the Sculpture Copyright Acts) and in paintings, drawings and photographs (under the Fine Arts Copyright Act 1862). Rights in designs of articles of manufacture were acquired by registration from 1839 on. The domains of these various forms of right – the 'aesthetic' and the 'applied' – were clearly distinguished from each other in the relevant legislation, reflecting the established philosophy that art and manufacture were quite separate from each other. The 1911 Act, however, opened up a bridge between the two by recognising that what it called 'works of artistic craftsmanship' might be the subject of copyright. Here the influence of William Morris and the Arts and Crafts Movement is very apparent, as Lord Simon of Glaisdale pointed out in the House of Lords case, *George Hensher Ltd v Restawhile Upholstery (Lancashire) Ltd* [1976] AC 64. Morris had asserted the need to combine artistry and craftsmanship in the production of goods, but had rejected the possibility that this could come about through the processes of mass production, emphasising instead the importance of handicraft by the individual artist-craftsman. So the bridge created between art and manufacture by the concept of 'artistic craftsmanship' was a narrow one indeed, given that the great majority of goods in circulation were mass-produced by machines.

But the 1911 Act also included drawings amongst artistic works, irrespective of their artistic merit; conferred an automatic copyright term of the author's lifetime plus fifty years; and provided that infringement might be by reproduction in any material form. Thus design drawings of all types could be the subject of copyright, a right lasting much longer than the registered design right, and one which could be infringed by the unauthorised production of industrial objects throughout the period of protection. The perception of a division between the terrain of copyright, essentially connected with art, and that of registered designs, tied to manufacture, nonetheless remained firm. It was to avoid cross-over that section 22 of the 1911 Act was passed. It provided:

> This Act shall not apply to designs capable of being registered under the Patents and Designs Act 1907 [*the then-prevailing legislation on registered designs*], except designs which, though capable of being so registered, are not used or intended to be used as models or patterns to be multiplied by any industrial process.

The aim of the Act here is manifest. Industrial designs were not to have the protection of artistic copyright. Registration was to be the means of protection. Unregistered designs could not claim copyright if they were capable of registration and were used or intended to be used for the production of manufactured articles. Too many difficult questions were left unanswered, however. What was meant by the phrase 'capable of registration'? What was the position of a design *incapable* of registration? The design which had not completed the registration process had no protection. Copying of all sorts was possible with unregistered designs so that, as Jeremy Phillips (1986: 266) put it, "an unauthorised use for purposes which were not connected with the industrial multiplication of goods bearing the design (for example, the making of greeting cards displaying the design in question) could not be prevented". A further difficulty, that the section had not succeeded in its basic policy of excluding copyright from the field of industrial designs, emerged in the Popeye case, *King Features Syndicate Inc v O M Kleeman Ltd* [1941] AC 417. Here it was held that copyright subsisted in drawings of the cartoon figure, Popeye the Sailorman, and that it had been infringed by the unlicensed production of Popeye dolls, toys and brooches.

The 1956 Act made a new attempt to exclude copyright from industrial designs, taking account of the problems which had arisen under the 1911 provisions and giving effect to the recommendations of the Gregory Report on Copyright Law (1952: Part X). A change of approach was adopted. Section 10 recognised the existence of copyright in industrial designs but restricted the ways in which that copyright could be infringed. First, where a design had been registered its protection from unauthorised reproduction in the industrial field depended entirely on the rights conferred by the Registered Designs Act. If a design was applied industrially (that is to say, to more than fifty articles) but not registered, it was to lose copyright protection in the industrial field against those acts which would have been infringements of the rights conferred had it been registered. In other words a design applied industrially but not registered had no protection against industrial copying until it was registered.

The effect of section 10 on the unregistered design which had been applied industrially was first judicially considered in *Dorling v Honnor Marine* [1965] Ch 1, in which the Court of Appeal held that, despite the section, such a design might have full artistic copyright where it was not registrable because it consisted of features dictated solely by the function to be performed by the article to which it was applied. The justifications for this decision were twofold. One was the correct observation that all designs had artistic copyright under the 1956 Act; section 10 only limited the ways in which the copyright might be infringed. Secondly, these limitations were to be worked out by examination of the rights arising had the design been registered. But the design in *Dorling* could not have been registered. To say that section 10 operated to deprive such a design of copyright protection had the effect that it could not be protected at all. The old maxim, "what is worth copying is worth protecting", applied; it had to follow that the copyright in an unregistrable design was not limited by section 10.

## Problems of Design Copyright

*(a) Term*

The result of the reasoning in *Dorling* was strange indeed when set against the general policy of restricting the nature and length of protection given to industrial designs against unauthorised reproduction. A registered design had protection for a maximum period of fifteen years; a registrable but unregistered design had no protection at all against industrial copying; and an unregistrable design had protection for the author's lifetime plus fifty years. A particularly odd result was that functional designs, which it was the policy of the Registered Designs Act not to protect at all, had achieved more protection than any registrable design by virtue of copyright.

The anomaly was reinforced by the House of Lords in the *Amp* case in 1972. As already discussed (above, 29-30), the effect of that decision was greatly to increase the number of designs which were unregistrable because they lacked eye appeal and were dictated solely by function in the sense worked out by the court. But, standing the correctness of the *Dorling* case, all such unregistrable designs had the much more powerful protection of copyright. In 1972, however, it might have been thought questionable whether *Dorling* was still good law. This was a possible argument following the enactment of the Design Copyright Act 1968, which substantially amended section 10 of the 1956 Act. The 1968 Act was the outcome of the criticism of the registration system voiced by the Johnston Report on Industrial Designs in 1962. The committee had attacked the lack of protection for unregistered designs which, it argued, was having unfortunate results given the cumbersome nature of the registration process. Many manufacturers found it too difficult and expensive to register their designs. The report proposed that unregistered designs should enjoy copyright protection, which would arise automatically with the creation of the design, but it should subsist only for the same period as the right under a registered design. Of course, in making this proposal, the committee had not had any opportunity to consider the *Dorling* case, which still lay in the future; regrettably and without having the same excuse, Parliament seems not to have considered the implications of the case in giving effect to the Johnston recommendations.

The 1968 Act conferred full artistic copyright on designs but only for a period of fifteen years from the date on which the article to which the design was applied was first sold, let for hire or offered for sale or hire. Thereafter no act in the industrial field could infringe the copyright in the design. The result was that a registered design now had dual protection for fifteen years under both the registration system and copyright, while the unregistered design had copyright protection for the same period. One might have thought that this would have resolved the unfairness identified in *Dorling*, the vulnerability of the unregistered design to appropriation by others, so that the interpretation of section 10 as applying only to registrable designs could have been abandoned.

There was still a technical difficulty, however. Since, like the original section 10, the 1968 Act made no reference to the question of the registrability of the design, it was possible to argue that it was not intended to change the law declared in *Dorling*. Accordingly, the new provisions did indeed apply only to designs capable of registration; if a design was not registrable, *Dorling* still stood and it had ordinary unlimited artistic copyright.

The question did not arise squarely in litigation for several years. In *LB Plastics v Swish Products Ltd* [1979] RPC 551 the House of Lords did confirm the copyright protection of a functional design without becoming involved in a discussion of the duration of the copyright; the main issues in the case were what constituted copying and the defences available under the 1956 Act. By this time, however, a flood of other cases was coming forward for decision. Many of them concerned parts of complex machines, in particular motor vehicles. Until the 1970s car manufacturers seem to have been little concerned by the activities of other manufacturers and dealers specialising in the supply of replacement parts for their vehicles. (For explanations of this, see below, 41.) The development of the law of copyright in relation to functional designs just outlined, however, appears to have been seen as providing a way of taking control of a thriving market at a time when profits from traditional activities were dropping for a variety of reasons. It seemed to give car manufacturers an exclusive right to reproduce the design work for their products, which would endure for far longer than the commercial life of those products or any patent protection which the product might have. As a result, the manufacturer had a choice: either he could exploit the market by himself, eliminating competition by copyright infringement actions, or he could regulate it by means of the threat of litigation against other manufacturers and the grant of licences on his own terms and conditions. Not surprisingly, these moves were strongly resisted by the spare parts industry and it was this that led to a spate of litigation in the early 1980s.

The most significant of these cases, because it went to the House of Lords, was *British Leyland v Armstrong Patents* [1986] AC 577. The facts were commonplace. The design in question was that applied to the exhaust pipes of Morris Marina cars, vehicles manufactured by British Leyland. It was an unregistrable design. Armstrong Patents manufactured replacement exhaust pipes for Marinas. By a majority of 4:1, the House of Lords confirmed the general position that unregistrable designs enjoyed full artistic copyright. The speeches show that all members of the panel were fully aware that somehow the law had taken a wrong turning but they were inclined to blame Parliament rather than to impugn the judicial interpretation of the 1956 Act and its amendments.

The result of this development of the law was quite extraordinary: a total reversal of legislative policy accomplished by judicial decision. The policy in general terms was to protect only designs with 'eye appeal', and even those only for a maximum period of fifteen years. Now all designs had protection if within the scope of the Copyright Act; those which it had been the legislative policy to exclude from protection altogether had it for a period a great deal more than three times longer than that provided by design legislation. Moreover, as spare parts cases made clear, design copyright, while not conferring a monopoly in the manner of a registered design, enabled manufacturers to take control of the market in their products and to eliminate competition if they so wished.

*(b) The Concept of Copying*

As we have seen, one of the features of the law which prevents copyright from being a 'monopoly' right like that of a registered design or trade mark of a patent is that infringement generally requires a process of copying: the infringing article must be derived from the original work in question. In the design copyright cases, however, this concept was stretched considerably by the courts so that the copyright conferred by the decision in *Dorling* was given full effect against competing products. This was in fact the main copyright issue dealt with by the House of Lords in *LB Plastics* and *British Leyland*.

The copying problem arose as a result of a variety of factors often found in design cases. First, in the reported cases it seldom occurred that the alleged infringer had actually copied or even had access to the original design

## THE RISE AND FALL OF DESIGN COPYRIGHT

documents. Instead a process known as 'reverse engineering' had been used, where the copyist had worked back from the designer's ultimate product to achieve the offending new product. Second, this was often linked with efforts at 're-originating' or 'redesign' where a designer was presented with a brief based on the original product and asked to produce a new design from the brief rather than the product; indeed, in what is known as a 'clean room' procedure, the designer might be kept ignorant of the product during his work. Third, many design drawings are not simply drawings but also incorporate what, in Copyright Act language, is literary material, that is, explanatory words and figures showing scales and dimensions.

These factors raised a series of problems for artistic copyright law. It was well understood before the *Dorling* case that indirect copying – copying copies of the original – constituted infringement of copyright. An example was the *Popeye* case, where the infringer copied, not the drawings, but the licensed dolls and brooches. It was relatively easy to extend this concept to 'reverse engineering' and this was done in *Dorling, British Northrop v Texteam* [1974] RPC 57, and *Solar Thomson Engineering v Barton* [1977] RPC 537, before the final confirmation by the House of Lords in *LB Plastics v Swish Engineering* [1979] RPC 551. The main difficulty here was that as a general rule the 'reverse engineer' had never seen the original drawings which he was supposed to be copying and, at least in visual terms, it was often difficult to say that copying the product was also copying the drawings. Where the reverse engineering was carried out through redesign, it became even harder to see the element of copying involved, even allowing for the validity of the concept of indirectness. Arguments that redesign was not copying sometimes succeeded (*Merlet v Mothercare* [1986] RPC 115; *Rose Plastics GmbH v Wm Beckett & Co* [1989] FSR 113). Nevertheless the courts tended to hold that there was still a causal link between the original design and the ultimate 'redesigned' product, the decision being perhaps made easier by the close similarity between the rival products at the end of the process. (See the *House of Spring* cases: [1983] FSR 213; [1985] FSR 327; [1986] FSR 63; and *Howard Clark v David Allan & Co Ltd* 1987 SLT 271.)

Perhaps the most difficult issue, however, concerned the literary element in design drawings. The alleged infringement consisted of making a three-dimensional reproduction of a two-dimensional work. The 1956 Act specifically provided that the copyright in a two-dimensional artistic work might be infringed in this way, but said nothing about literary works in this respect, apart from the fact that infringement might be by reproduction in any material form. In *Brigid Foley v Ellott* [1982] RPC 433 it was held that to manufacture garments was not reproduction of the words and numerals constituting a knitting guide. A reproduction, according to Sir Robert Megarry VC, must be "some copy of or representation of the original". In the *Interlego* case it was held that to produce an article by following written instructions was not an infringement of the copyright in those instructions. Whitford J made the following comment in *J & S Davis (Holdings) Ltd v Wright Health Group* [1988] RPC 403, a case about dental impression trays:

It may no doubt be true that you do not infringe copyright in a recipe by making a cake. If the defendants had made, from a table of measurements, trays which were in general accord to the teaching of Professor MacGregor, no case of infringement would have arisen. There is no three-dimensional provision applicable to literary works.

Further, in many engineering drawings the artistic element is relatively insignificant, being meaningless without the associated literary matter, and so what is being copied is that literary matter; but the copying falls outwith the scope of reproduction in respect of literary works. Such an argument can be supported from cases. In *Duriron v Jennings* [1984] FSR 1, for example, it was held that there was no artistic copyright in a table of statistics attached to a rudimentary engineering drawing which was not to scale. But Dillon LJ distinguished this from the case of a detailed manufacturing drawing where the artistic copyright would cover the drawing interpreted in the light of writing and figures upon it. This was also the view of Oliver LJ in *British Leyland v Armstrong Patents*. In *Anacon Corporation Ltd v Environmental Research Technology Ltd* [1994] FSR 659 it was held that a circuit diagram was both a literary and an artistic work, although it was not necessary for the judge to say whether the literary copyright was infringed by the defendant's three-dimensional circuit board. However, one could argue that the point was anticipated and met by the definition of 'drawing' in section 48(1) of the 1956 Act (now, as 'graphic work', under s 4(2) of the 1988 Act) as including maps, charts, diagrams and plans – all documents which are bound to include written material, but which nonetheless were brought under the heading of artistic work. It could further be suggested that there is no need to confine the material form required for infringing reproduction of a literary work to two dimensions, although this would be more difficult, especially after the judicial observations quoted above. Since the point is specifically dealt with in relation to artistic works, it is arguable from silence that under the Act it is not possible to have such infringement in relation to other forms of work.

Taken all in all, however, the design cases did show some stretching of the traditional concepts of copying in order to enable action to be taken by the copyright owner. In *British Leyland v Armstrong Patents* an assault was mounted on the concept of indirect copying and in particular on the *LB Plastics* case. This involved an invitation to the House of Lords to exercise its power to over-rule its own previous decisions, a power which is relatively rarely exercised. Only Lord Griffiths was willing to do so; the other Law Lords apparently regarded the precedent as correctly decided and capable of change only at the hands of the legislature, which alone could properly take account of all the issues involved. As Lord Bridge remarked, "the line of authority to which the House gave its imprimatur in the *Swish* case has now been followed for more than 20 years and must have had a profound impact on many areas of industrial practice" ([1986] AC at 623).

*(c) Form of Design*

Copyright was clearly an attractive form of product protection to many manufacturers. It was ready of access; to acquire copyright cost nothing. It was long-lasting, more than capable of covering typical product lifespans, and powerful remedies were available to bring to bear against those seeking to compete without licence. Yet it was by no means a perfect form of protection. It was fundamental to most of the cases that the copyright infringed lay in drawings. But not all design work was founded on drawings. This was illustrated in *George Hensher v Restawile Upholstery* [1976] AC 64, where designers of suites of furniture worked from three-dimensional prototypes. Copyright was claimed in these prototypes as works of artistic craftsmanship in order to challenge the products of a rival manufacturer, but the House of Lords rebuffed the argument. The increasing use of computer technology in industrial design also caused difficulties for a traditional copyright approach. It is obvious that a person composing a computer program is an author and (subject to the rules about employment) first owner of the copyright in the program, but where as part of its operations the computer itself produces works it is less easy to see the programmer or the operator as the author. In *Express Newspapers v Liverpool Daily Post and Echo* [1985] FSR 306 it was held that the copyright in a work produced with the aid of a computer pertained to the programmers who wrote the software used in the operation. The Copyright (Computer Software) Amendment Act 1985 made it clear in section 2 that a work held on computer had been recorded in a permanent form so that it could claim copyright. But this did not deal with all the possible issues capable of arising from the use of computers in design work, in particular the situation where the computer generated the design without any significant human intervention. On the other hand, copyright in functional designs was beneficial to the computer industry, since it was suggested that the law thereby protected the topography of semiconductor chips. The importance of this argument was that it enabled British producers to persuade the US Government to grant them protection against copying in the United States under the Semiconductor Chip Protection Act 1984.

## Challenging Design Copyright

Despite its obvious inconsistency with the policy of the law manifested under not only the Registered Designs and Patents Acts but also the Copyright Acts themselves, copyright became firmly established during the 1970s and 1980s as a means of protecting products from competition. The *Interlego* case, in which the Lego company argued that its own registered design was unregistrable in order to claim the benefit of copyright instead, thereby closing out potential competitors for a much longer period of time, showed that the registration system was being ludicrously distorted by design copyright, even though the argument was ultimately unsuccessful. At least in part design copyright developed because the design registration system, already complex, became more difficult as a result of the *Amp* decision in 1972. But it was already clear, from the Johnston Report (1962) and the Design Copyright Act 1968, that the system was not serving certain sectors of industry particularly well. While one might criticise the incursion of copyright into the field, it had to be recognised that the development was itself an implied criticism of the registration system. At the same time, the expansion of copyright showed the strength of the idea that the originator of a work involving some degree of skill and labour, such as product design, had a right of property in the result and that copyists were encroaching upon these rights and misappropriating the results of others' work.

The development of design copyright did not go unchallenged. Various avenues of attack available under the existing law were explored in the courts and elsewhere, in an effort directed to at least restricting the impact of copyright, if not of excluding it altogether. This was most significant in relation to spare parts and the motor industry. One possibility was a rather opaque provision of the Copyright Act 1956; the second, recognising that at root many of the problems were about the ways in which particular markets should operate, was through competition law, first of the United Kingdom, then of the EC. Last, and most effective in the short term, was to rely on the common law – or, as it has been commonly described and in this case would have been very aptly described, the judge-made law. We shall take each in turn, some more briefly than others.

*(i) Section 9(8) of the Copyright Act 1956*

Section 9(8) of the 1956 Act provided that the making of an object of any description which was in three dimensions should not be taken to infringe the copyright in an artistic work in two dimensions, if the object would not have appeared to be a reproduction of the artistic work to persons not experts in relation to objects of that description. It is not necessary here to go into the remarkable judicial nullification of this provision, which was intended to restrict the enforcement of artistic copyright against three-dimensional reproductions. Full details may be found in the first edition of this book (MacQueen 1989: 39-42). Suffice it to note that the cases on the provision demonstrated a consistent determination not to apply it, often justified by the obscurity of meaning which arose from the use of a triple negative in the sub-section coupled with judicial distaste for the idea that a copyist who would not be detected as such by a non-expert was not an infringer. Once again the deeper questions of policy were overlooked; although it had to be said that the legislative draftsman had failed to make clear the relevant considerations.

*(ii) Domestic competition law*

A second avenue of attack on design copyright and its exploitation in an industrial setting was through British competition law. In May 1984 a reference in the following terms was made to the Monopolies Commission under the Competition Act 1980:

(a) The persons whose activities are to be investigated by the Commission is Ford [Motor Company Limited];

(b) the goods to which the investigation is to extend are replacement body parts;

(c) the course of conduct to be investigated is the pursuit by Ford of a policy and practice of not granting to any person (other than to persons supplying body parts to Ford) a licence to manufacture or sell in the United Kingdom any replacement body part where Ford claims to be entitled to prohibit such manufacture or sale by virtue of the copyright subsisting in the drawings or designs of body parts.

For the purposes of this reference:

'body part' means any body panel fitted to a motor vehicle as standard equipment when sold new and 'replacement body part' means any body panel sold as complete or partial replacement for a body part; 'copyright' shall include copyright arising from the registration of a design under the Registered Designs Act 1949.

The Commission reported in February 1985. The report outlined the nature of the market in replacement panels for Ford vehicles. The market was served by Ford itself and a group of manufacturers described as the 'independents'. The total value of the market was between £35 million and £45 million per year, with about 85% of that being taken by Ford. It appeared that Ford had been indifferent to the activities of the independents so long as these were confined to 'corrosion part panels' – that is to say, small parts of body panels "intended to replace those areas of a car body most susceptible to rust, being welded into position when the affected part has been cut away". The sales of such panels became significant following the introduction of the MOT test in 1967. "It was only when independents started to supply full panels, which constituted a threat to a significant part of its spare parts business... that [Ford] felt compelled to initiate legal proceedings" and indeed to move into the supply of corrosion part panels. Ford had also been registering the designs of its full panels since November 1980. The company "acknowledged that the success of its legal actions would eliminate the competition provided by independent suppliers of panels which infringed Ford's copyright or registered designs but considered such competition to be unlawful, in the absence of any licence granted by Ford under its copyright or registered designs". It was unwilling to grant licences to independents except for corrosion part panels.

The Commission concluded that Ford's refusal to grant licences in respect of its body panels was an anti-competitive practice, since it had the effect of preventing competition in the supply of body panels in the United Kingdom. Moreover, the practice was adverse to the public interest. Ford's prices were higher than those of the independents and the competition tended to keep the Ford price down. Corrosion part panels had been introduced by the independents and clearly satisfied a previously unmet need. It was doubtful whether Ford would have done this. "In this way, competition has been the spur of innovation and has brought benefit to members of the car-owning public." Ford argued that the competition of the independents was unfair since their low prices were the result of not having to pay for either research and development or a royalty, pointing out that over the five years from 1979 to 1983 the company had spent £377.6 million on R & D. There had to be some return on this level of investment. The incentive for Ford to keep its replacement part prices low was that the consumer took the cost of obtaining spares into account in making his initial purchase. The Commission's reply to this line of argument was, first, that Ford's sales revenue from replacement parts (between £30 and £40 million in 1983) was only a small portion of its total sales in the United Kingdom (£2,679 million in 1983); and second, following survey evidence, that the cost of replacement parts was not a crucial factor influencing the decision of purchasers, and was accordingly not a sufficient control on Ford's freedom to set prices in a market otherwise without competition.

A number of other arguments were considered by the Commission in deciding that Ford's licensing policy was against the public interest. It did not seem unduly impressed by the suggestion that the independents' parts were of poorer quality and less safe, or by the argument that employment at Ford's would be reduced: the boot, if anything, was on the other foot.

The difficulty lay in providing a remedy for Ford's anti-competitive conduct. The Secretary of State had no power under the Competition Act to order the grant of licences by Ford. Some change was therefore necessary in the law which gave Ford its capacity for acting in an anti-competitive manner. The Report went on to make various recommendations for law reform, which will be considered in the next chapter, and expressed the hope that the problem would be resolved meantime by the parties involved.

*(iii) European Competition Law*

This was not the end of the Ford case. In 1985 the European Commission commenced an investigation into the matter under Community competition law. As already noted (above, 20), the case was settled in 1990 rather than pressed to a conclusion, Ford having given undertakings that it would not insist upon absolute exclusivity in the exercise of its rights and that it would grant licences on reasonable terms ([1990] 5 EIPR D-101). Even before the reference was made, questions had been asked in the European Parliament about Ford's policy as a possible infringement of Article 86 of the EC Treaty, that is, as an abuse of a dominant position in the common market. Commission answers suggested that Article 86 might apply if the consumer was, as a result of the use of copyright law, compelled to depend on 'captive parts' produced by the car manufacturer ([1983] 4 *European Competition Law Review* 138). Another indicator of the Commission's view of design copyright was a passage in the Green Paper on copyright (1988: paras 1.3.5-6), to the effect that the "restrictive effects of copyright protection on legitimate competition have on occasion risked becoming excessive, for example, in respect of purely functional industrial designs". Copyright protection should be limited, for otherwise there will be a "genuine monopoly, unduly broad in scope and lengthy in duration".

The direct enforceability of Community competition rules in national courts led to their being invoked in litigation over design copyright in the United

Kingdom. Initially, and especially in the spare parts cases, such arguments had some success, although there was always concern that the EC Treaty should not become a pirates' charter (MacQueen 1989: 44-5). A number of cases involved British Leyland, which by contrast with Ford was prepared to allow others to manufacture and supply its spare parts but only if licences were taken. This was challenged as in itself an abuse of dominant position: Leyland's alleged status as the largest motor manufacturer in Europe, coupled with a copyright in spare parts, gave the company a dominant position, the abuse of which was its effort to control competition through the exercise of copyright. The rejection of this argument came in the Court of Appeal judgment in *British Leyland v Armstrong Patents Co Ltd* [1986] RPC 279. There Oliver LJ held that 'dominant position' was an economic concept ([1986] RPC at 304-7):

> No doubt the owner of any industrial property right which enables him to monopolise production or reproduction may be said, in one sense, to 'dominate' those who wish to make use of his monopoly...[but] The owner of a monopoly right does not, by that fact alone, enjoy a dominant position simply because he can, if he wishes, refuse to grant licences. Dominant position depends upon the facts of the individual case, and whether as a matter of fact the allegedly dominant concern is able practically to behave independently of its competitors, purchasers and suppliers without taking into account their interests to any substantial extent – for instance, where it can determine the prices without reference to competition or control production or distribution of a significant part of the relevant goods.

Since it had been found that Leyland had only 24% of the market while their licensees had 36%, and since there was evidence that in pricing Leyland had to respond to the competitive nature of the market, there was no dominant position even though there might be a monopoly flowing from the copyright. Even if there was a dominant position, there was no abuse: there were no complaints from licensees, and the terms of the licences and the royalties charged by Leyland for them were not unreasonable.

Article 85 complaints principally concerned the terms agreed for licences by copyright owners and their licensees. In *British Leyland v Armstrong Patents Co Ltd* Oliver LJ thought that there was "a very short answer to the defence based on this Article" ([1986] RPC at 303). The licence agreements might or might not be void, but that made no difference to the existence of British Leyland's copyright and its general enforcement against those not parties to a licence agreement, such as Armstrong Patents.

The judgment of the Court of Appeal on the 'Euro defences' issue in *British Leyland v Armstrong Patents* was very strongly in the negative. While it was not said that such defences could never be taken against the exercise of design copyright, it suggested that they would be extremely difficult to make out. Perhaps significantly, the House of Lords did not discuss the issue on appeal.

One of the difficulties confronted by the British courts in considering the possible application of the EC Treaty to design copyright was the absence of any definitive rulings on the matter from either the European Court or the European Commission. In *Keurkoop v Nancy Kean* [1982] ECR 2853 the Court held that a design registration system (in this case, the Benelux system) was 'industrial and commercial property', the rights in which might be exercised in derogation from the free movement of goods within the Community, provided that the goods in question had not already been put on the market in the exporting Member State by or with the consent of the right-holder. The decision was of little guidance, however, on the key issue of design rights in spare parts. The Commission had said in the *Hugin* case ([1978] 1 CMLR D19) that a company might enjoy a dominant position in respect of the supply of spare parts for its principal product where it was the only source, and that ceasing or refusing to supply parts might be abuse of that position; and this had been upheld by the Court ([1979] ECR 1869). But that case did not involve intellectual property rights. (Note also more recently *Hilti AG v Commission* [1994] FSR 760). There were a number of other indications, however, that the Commission disliked the deployment of intellectual property rights to tie customers into buying spare parts for the right-holder's product, or to prevent licensees and others from providing them. Thus in the Block Exemption for Franchise Agreements it is stated that while a franchisor may prevent the franchisee from selling a competitor's goods, the latter cannot be prevented from manufacturing, selling or using competing spare parts (EC Commission Regulation 4087/88, article 2(e)). Slightly less directly, the Trade Mark Directive and Regulation both provide that it is not infringement of a trade mark to use a mark in the course of trade where it is necessary to indicate that the purpose of a product or service is to be an accessory or spare part, provided that this is in accordance with honest practices in industrial or commercial matters (Council Directive 89/104/EEC, art 6(1)(c); Council Regulation (EC) No 40/94, art 12(c); and see Trade Marks Act 1994 s 11(2)(c)).

It was late 1988 before the Court was able to make a ruling on intellectual property rights in spare parts and then again it was in the context of design registration systems. There were in fact two cases, one from Italy and the other from the United Kingdom, both concerning replacement parts for cars. In both the European Court held that it was not in itself an infringement of Community law to use the rights conferred by design registration to prevent the manufacture and sale of protected products by unauthorised third parties.

The first of the cases was *Consorzio Italiano della Componentistica di Ricambio per Autoveicoli v Regie Nationale des Usines Renault* [1988] ECR 6039. Under Italian law Renault held design patents for car body parts, which were being challenged by an Italian association of motor vehicle spare parts suppliers. The Court, applying the *Keurkoop* decision, held that under Article 36 Renault were entitled to use their rights under Italian law. It went on to find that there was no abuse of a dominant position under Article 86 merely because a party had obtained the benefit of an exclusive right which enabled him to prevent others from manufacturing and selling goods. There might be abuse where the party arbitrarily refused to deliver spare parts to independent repairers, or charged excessively high prices, or decided to cease production of certain parts when many vehicles of the design were still in use. The mere fact that the original manufacturer's prices were higher than those of independent manufacturers was not necessarily an abuse, however, "because the holder of an ornamental design right can legitimately claim remuneration for the expenses it incurred to perfect the protected product".

The judgment of the Court in the United Kingdom case of *AB Volvo v Erik Veng (UK) Ltd* [1988] ECR 6211 was along similar lines. The subject matter of the dispute was a registered design for the front wings of Volvo series 200 cars. Veng had been importing panels, manufactured without Volvo's consent, for sale in the United Kingdom. Volvo had refused to grant licences for spare parts. Applying once more the idea that the right-holder is entitled to exercise his rights, the Court ruled that to impose upon him an obligation to grant licences to third parties, even in return for a reasonable royalty, would be to deprive him of the substance of his right. Refusal to grant licences was not therefore in itself abuse of a dominant position. That would occur in the circumstances outlined in the Renault case.

These judgments confirmed the correctness of the approach taken by Oliver LJ to 'Euro defences' in *British Leyland v Armstrong Patents*, and were entirely in line with the balance which Community law in general has sought to establish between legitimate and illegitimate uses of intellectual property rights. The decisions that refusal to grant licences is not necessarily abuse of a dominant position had interesting implications for the Ford case, but it was still the case that other factors there may have made Ford's policy abusive (Laddie Prescott & Vitoria 1995: para 17.41). Similarly, the statement in the Renault judgment that merely charging a higher price than an independent is not an abuse and may be a justified exercise of rights has important implications in considering the monopoly conferred by intellectual property generally. Taken along with the Volvo decision, the cases indicated that the answer to the problem of spare part monopolies arising through intellectual property was not to be sought through European competition law.[2] On the other hand, it can also be observed that the Commission's challenges via competition law was to a very specific area in which it thought intellectual property rights were inapt, and that they served the purpose of bringing these rights under review for purposes of legislative reform.

## (iv) Licence to Repair: No Derogation From Grant

The *British Leyland* case in 1986 has already been referred to a number of times in the course of this chapter. It represented the highwater mark of the development of design copyright in the United Kingdom, being the most authoritative declaration of all that unlimited artistic copyright subsisted in functional design work. But the case also saw the turn of the tide, in that, despite its findings as to the general position on copyright, the House of Lords found for Armstrong and dismissed British Leyland's action. The House was manifestly conscious that the law under its interpretation of the 1956 Act was unsatisfactory from the point of view of policy, and accordingly set out to limit its effect by drawing on other principles of law found outside the legislation.

There was already a thread of authority holding that the owner of goods held an implied licence to infringe copyright in repairing them, drawn from the existence of such a principle in patent law. It appeared, however, that the right of repair did not permit third parties to set up a business of manufacturing and supplying replacement parts without a prior specific order from the owner. This thread was picked up in the *British Leyland* case, in which it was held at first instance, and again in the Court of Appeal, that Armstrong could not plead a licence to repair, because they were third parties whose business of making and supplying spare parts arose before any order or commission from owners of goods in which there existed design copyright. The House of Lords, however, set the whole argument on a new footing by shifting away from the concept of implied licence to a consideration of the rights of an owner of a product. It was clear that in general there was a right to repair what one owned and there was no need to invoke implied licences to create it. The questions were, therefore, how far did this right of the owner extend in enabling him to contract with third parties for repair, and to what extent was his right limited by the rights of others?

The court answered these questions by applying the common law principle against derogation from grant – that is to say, one is not allowed to give a thing with one hand and take it back with the other. The principle is familiar in both Scots and English law and has been used mainly in certain well-defined relationships, such as those between landlord and tenant, or the vendor and purchaser of a business. Nonetheless it seems clear that the principle is one of general application and one can readily imagine situations involving copyright where it could operate: for example, the granting of copyright licences. Nor is it confined to contractual situations: a common example of the application of the principle is to rights of servitudes and easements, which do not arise simply from contracts between the parties but also from their relationship, and the relationship of their respective successors in title, as proprietors and users of land.

Where was the grant in the *British Leyland* case? The speeches of the majority suggest that it consisted in the manufacture and sale of the cars to the public: the beneficiaries were members of the public acquiring the vehicles. Because car manufacturers seldom if ever sell direct to the public, the grant was not a contractual one. Moreover the grant benefited not only the first owner but also all subsequent ones. While there was a vague parallel with easements and servitudes, the application of the principle here went well beyond any previous case. There is clearly an on-going legal relationship between the owners of neighbouring properties, which is quite distinct from that between the owner of a second-hand car and its manufacturer.

Further, the defendants in the case, Armstrong Patents, were not the beneficiaries of the grant and, on the orthodox view, as strangers to it were not entitled to plead the non-derogation principle. The argument of the majority in response to this point was that the activities of third parties such as Armstrong were the only way in which the beneficiaries of the grant could economically exercise their right to repair their property. A simple answer to such an argument, however, is that it is not the function of the law, or of the principle against derogation from grant, to enable people to exercise their rights more cheaply, particularly when it is done at the expense of the rights of other people. Moreover, this kind of reasoning could be extended to justify the activities of copyright pirates, because they enable the public to exercise the right to listen to music or view films more cheaply than is possible from the versions made available through the copyright owners. Further, unlike Ford – an example referred to by Lord Bridge – British Leyland were not proposing to prevent the public from exercising the right to repair through third parties. This was why the court emphasised that derogation occured as a result of British Leyland's control of the market leading to an uncompetitive situation where the lowest possible price would not be obtainable.

Lord Griffiths, dissenting, made some apposite remarks in his speech:

> Whenever Parliament grants a right in the nature of a monopoly, and copyright is such a right, it inevitably interferes in some respects with the freedoms of the public that would exist apart from the monopoly... If Parliament has, through copyright, given a monopolistic right to the manufacturer in the shape of his spare parts, on what principle is the court free to refuse to enforce that right given by Parliament to the manufacturer?... It seems to me highly improbable that a motor car manufacturer would exploit his copyright either to starve the spare parts market or to increase the fair price for his spare parts for I can think of nothing more damaging to his prospects of selling the car in the first place.

It is the argument of this book that the development of the law giving unlimited artistic copyright to functional industrial designs was mistaken, but it does not follow that the protection of such designs was or is inappropriate. The difficulty with the derogation from grant principle is its scope. In the *British Leyland* case it operated to render a copyright nugatory. This was a drastic conclusion and one the scope of which for copyright law generally was and is quite unquantifiable. Clearly it was aimed very specifically at the facts of the case and was an attempt to reorient the law in accordance with the basic policy on industrial designs, as indicated by Lord Templeman's remark that "the exploitation of copyright law for purposes which were not intended has gone far enough" ([1986] AC at 644). The House was confronted with a situation where the law seemed to leave, in Lord Bridge's words, "no halfway house" ([1986] AC at 626). There was no mechanism for assessing the reasonableness or otherwise of British Leyland's licensing policy or of monitoring its administration subsequent to any court decision in the company's favour. Finding a solution in the combined operation of the Copyright Acts and the common law did avoid the need to investigate the *terra incognita* of Article 86 of EC Treaty, which in any event Oliver LJ had demonstrated to be inapplicable. But the answer provided by the House was inflexible and not well adjusted to meet the difficulties of the whole issue of design protection. It is also now evident, in the light of the Renault and Volvo cases, that the House went much further than the European Court would have been prepared to do, by depriving British Leyland of the substance of its intellectual property rights. The House knowingly went further than the Monopolies Commission in its report on Ford, where limiting, rather than removing, protection from designs was, as will be seen in the next chapter, the main recommendation. The unease of the court is evident in the calls by members of the majority for new legislation. That, as they knew, was on its way; we shall turn now to consider how that legislation, when it came, was formed.

# Chapter Three

## The Road to Reform

The problems of design copyright and its interaction with the registered design system, outlined in the previous chapters, were evident by the early 1970s. In particular, the decision of the House of Lords in *Amp v Utilux* seemed to make it the only protection available for a very large number of designs which might previously have been registered. In August 1973 the Department of Trade announced the appointment of a committee:

> to consider and report whether any, and if so what, changes are desirable in the law relating to copyright as provided in particular by the Copyright Act 1956 and the Design Copyright Act 1968, including the desirability of retaining the system of protection of industrial designs provided by the Registered Designs Act 1949.

The committee was chaired by Mr Justice Whitford and its report was published in 1977. Chapter 3 of the Report dealt at length with industrial designs and made various recommendations. This was to spark off a decade of discussion, during which at least four other Government papers touched upon the matter, before a Bill was finally presented to Parliament during session 1987-88. The first official response to the Whitford Report was a Green Paper published in 1981; a further Green Paper appeared in 1983. Both papers contained substantial discussion of the law on industrial designs. 1985 saw the publication of the Monopolies Commission report on Ford, which made a number of recommendations on design law. In April 1986 the Government published a White Paper, with proposals for legislation. Lastly during session 1986-87 the House of Commons Trade and Industry Committee looked at the question as part of an investigation of the motor components industry.

Each of these reports made differing recommendations, and the scheme presented in the Bill was yet a further solution to the problem. The issues of law and policy involved changed little over the years: one of the striking features of the Whitford Report is its coverage of all the matters which were to vex the Government ten years later in drawing up and defending its Bill. Equally, it was not neglect which led to the delay in acting on the Whitford Report: rather, it was the struggle to find the right solution to a very complex issue. What is very evident in reading the published debate is a conflict of views on the matter within British industry. For some sectors the comprehensive protection of the existing law was of great value; for others it was a very serious inhibition on activity. The law contained a mixture of advantages and disadvantages from the point of view of small businesses, and this became an even more important issue with the accession in 1979 of a Conservative Government proclaiming its commitment to the creation of an environment in which such businesses could flourish. The issue also became entangled with the Government's policies of encouraging new enterprise and innovation in British industry and removing restrictions on and impediments to competition, all of which had to be achieved without placing foreign business at an advantage in the world market. All this took place against the background of the EC's own efforts to develop a single European market by removing barriers to trade and competition between Member States. Finally there was the problem of semiconductor topography protection. As already noted, the US Semiconductor Chip Protection Act 1984 required the rest of the world to take steps for the protection of semiconductor topographies, and by 1986 the EC had decided to respond with a Directive, which was eventually promulgated in December that year (87/54/EEC). The United Kingdom had claimed that appropriate provision was made under design copyright, so the EC response to the American demand for reciprocal protection was to be an important factor in shaping the new law on industrial designs generally.

This chapter seeks to illustrate the twists and turns of the debate by taking each of the principal issues in turn and considering the various approaches to each one found in the successive reports. This provides a setting for the discussion in the next chapter of the final resolution of the issues embodied in the Copyright Designs and Patents Act 1988.

## Protecting Functional Designs as well as Aesthetic Designs

The Whitford Report distinguished designs into two categories: those where the appearance of the article to which it was applied influenced the decision of a customer to purchase (Category A designs); and those where the customer's decision was not so influenced (Category B designs). The Committee was unanimous in the view that Category A designs should receive legal protection but was divided in respect of Category B designs. The majority were in favour of protection, applying the general principle that "what is worth copying is worth protecting" and that the investment, skill and labour involved in producing functional designs should be protected from competitive copying. "People are encouraged to produce their own developments and improvements and to enter into manufacture if they can be sure that the effort and not inconsiderable expense involved in putting a new product on the market will be adequately protected" (50). The Consumers Association had expressed particular concern about the anti-competitive effects of protecting spare parts, particularly for motor vehicles and domestic machinery. The Committee observed, however, that while competition held prices down, it did not appear to lower them. There was little evidence that copyright had curtailed the market in spare parts.

Two members of the Committee dissented, however, concluding that there should be no protection whatsoever. They argued that functional objects gained all the advantage necessary through being first on the market; legal protection would be against the public interest as it was likely, or could be used, to restrict competition. Attention was drawn to the problem of spare parts in an interesting passage (49):

> In the case of an article which has to fit, such as a battery for a radio, torch, or portable dictating machine, they [*the dissenting minority*] point out that this must of necessity be of a particular shape and that shape must be copied by any manufacturer, other than the original one, who wants to manufacture such a battery. Here again, they feel that a copyright protection could result in a monopoly, in that, if a manufacturer produced a radio with a novel shaped battery, he would *ipso facto* obtain a monopoly protection in that battery.

With functional objects what was important to the customer was their functional efficiency, so there was no case for protecting their appearance: otherwise there arose the danger of protecting the principle of the way the object worked when that would not attract patent protection. It would be a positive benefit to industry if there was a common pool of non-inventive ideas upon which all could draw.

The 1981 Green Paper, while accepting that there should be protection for designs with customer appeal, preferred the view and reasoning of the Whitford minority on functional designs (5-6):

> [T]he purely functional should not be protected against copying; indeed unless it attracts patent protection as being inventive it should not be protected at all... The Government does not believe that industrial progress will be helped if almost any industrial product is protected against copying. The Government thinks that if an industrial society is to be active and competitive, there must be a substantial common pool of experience from which all can freely take.

But the opposite conclusion was expressed in the 1983 Green Paper, although, as will be seen in the next section, the protection of functional designs was to be achieved by extending the scope of the registration system. The guiding philosophy of the 1983 Paper was the need to increase industrial awareness of, and access to, intellectual property rights with particular reference to "small but enterprising firms [which] are least able to utilize the present rights by may need them most to protect their main asset – an innovatory product" (1). The Paper saw greater clarity in the law as important, being concerned that the uncertainty as to the scope of protection for designs inhibited small companies from seeking to enforce their rights. Unfortunately the Paper did not squarely address the issue of whether functional designs should be protected at all but rather worked on the basis that they were, assuming rather than arguing that this should continue. As a result, it was not a particularly coherent contribution to the debate on this point.

Nonetheless when the White Paper was published in 1986, it was apparent that the Government had shifted from its 1981 position to a recognition of a need to offer protection to functional articles. The Paper stated (19):

> British industry relies to a great extent on its innovative abilities. It is clear that there are many innovative industrial products which are costly to design but which are not truly inventive and which therefore do not qualify for patent protection.

Accordingly, the Government has concluded that some protection should be available to give the manufacturer who has spent money on design the opportunity to benefit from his investment, thus providing an incentive to further investment. It has also concluded that this protection should extend to spare parts.

The Paper proposed the creation of a new form of protection "on copyright principles but without the more objectionable features of full copyright protection", the new right "to apply to both functional and aesthetic designs" (20-21).

This proposal met with a hail of criticism from one sector of industry, that of automotive parts manufacture and supply. The British Automotive Parts Promotion Council, which represented vehicle component manufacturers, distributors and 'quick-fit' high street operators, and the Industrial Copyright Reform Association launched an attack focused on the problem of the supply of spare parts for motor vehicles, the subject matter of the *British Leyland* case. The judgment of the House of Lords had seemed to free the spare parts market from the restrictions imposed by the copyright claims of original manufacturers; now the White Paper appeared to propose the resurrection of intellectual property constraints, albeit in a new form. The protests found a formal outlet in the Trade and Industry Committee of the House of Commons, which by coincidence was investigating the UK motor components industry during session 1986-87. Much of the evidence given to the Committee dealt with this matter.

The critique had two major thrusts. One was that the new right would be in favour of the vehicle manufacturers, which would have an adverse effect on the consumer. The Industrial Copyright Reform Association presented evidence suggesting that spare parts for DIY operations could be obtained on the high street at 42% of the price charged by the holder of a franchise from the original manufacturer, while the price from a 'fast fitter' would be 64% of the manufacturer's price. Yet there was no difference in the quality of the parts being supplied. This was because the manufacturers were often not the sole originators of the components used in the construction of their vehicles. Rather, the parts were bought in from the component manufacturers who were also seeking to supply the after-market. Vehicle manufacturers might more properly be described as assemblers of components manufactured by others, although the design of components was often a matter for close co-operation between vehicle and component manufacturers. Further, even where the vehicle manufacturer was the designer of the shape of the component, it did not necessarily follow that he should be able to claim sole rights in the result, as a representative of the engineering firm Turner & Newall explained to the Committee (Trade and Industry Select Committee Report 1987: 204):

> By and large if you take a component in a car that may have been originally designed by the original manufacturer and supplied by him, the fact that there is merely a design shape put on a piece of paper does not really represent the technology that is going into the individual product. In the case of many products it is the processing that is far more important than the actual piece of paper design.

## THE ROAD TO REFORM 53

This argument could of course be met with the suggestion that, if this were so, then the component manufacturers might benefit from the proposed new right rather than lose by it. Again, however, evidence was given to the effect that some vehicle manufacturers pressed component suppliers for transfers of intellectual property rights as a condition of obtaining contracts to supply them, and that by no means all were in a market position to resist such pressure. Further the components market was traditionally a free one and it was not in the interests of a component manufacturer to claim intellectual property rights, since that would invite retaliation by competitors.

The second argument against the White Paper proposal was the advantage which it would give foreign manufacturers, who would be able to prevent British companies supplying spare parts for their vehicles in Britain while not in a position to do so elsewhere. This was a potent argument for the Trade and Industry Committee, the main concern of which in instituting its inquiry had been the decline of the British industry, and the apparent success of overseas competitors in both home and foreign markets. The British Automotive Parts Council suggested that the White Paper proposals might, if carried into law, "force products to be made in European factories of our members rather than in British factories of our members in order to make sure they can put those products on the market very early in the life of a vehicle" (Trade and Industry Select Committee Report 1987: 289). The result could therefore be the loss of 400,000 jobs in British industry.

The Committee also took evidence on 6 May 1987 from Mr John Butcher, the Parliamentary Under-Secretary of State for Industry, whose responsibility it was to prepare the Bill to follow on the White Paper. It is apparent from his evidence that there had been further consideration of spare parts within the Department of Trade and Industry. His evidence included the following statement (Trade and Industry Select Committee Report 1987: 293):

> In the light of a number of representations that have been made to us by, in particular, the component suppliers to the after market, we believe that we should try to produce a formula which formalises the rights of those suppliers in a competitive regime to supply parts which must fit and must be a particular and very special shape, such as a gasket, completely without restrictions. In addition, we believe that we should formalise the position of, for example, a water pump which may be the design of the original equipment manufacturer but which an alternative supplier would like to fit on a vehicle with his own design but with a concession to that supplier that he can borrow from the original design those parts of it which allow it to connect to the car itself. On those two questions we believe we should provide clarity in order to preserve the competitive regime of the after market component suppliers. But also, I think, to allow, where exceptions must be made, the interests of the original equipment designer to get a proper return on his investments.

It is clear from a memorandum submitted to the Committee by the Industrial Copyright Reform Association that these 'must fit' and 'must match' exceptions to design right had been formulated in the Department by October 1986, that is, within a few months of the publication of the White Paper. The Association was sceptical of their practicability, and it is not clear from the

Committee's questioning of Mr Butcher that their import had been fully grasped by its members. The Committee's conclusions on the whole evidence presented to it was, however, clear enough. Its final report stated that the Government should ensure "that copyright protection does not extend to functional and mundane items, which includes most spare parts" (Trade and Industry Select Committee Report 1987: para 72). The Government response, dated November 1987, was to agree that there should be no monopolies in spare parts and that the provisions of the Bill then progressing through Parliament "will ensure that competition in the spare parts market is preserved" (HC 316, Session 1987-88, Trade and Industry Committee, First Report).

## Registration or Automatic Copyright

Given that in general design work deserved protection, how should that be achieved? British law had two schemes of protection, one depending on registration of the design, the other arising automatically whenever the design was rendered into permanent form as a drawing or other artistic work. The duality arose because not all designs, in particular functional designs, were registrable. Double protection was clumsy and inefficient. Which of the two models served the public interest better? Or were there other alternatives?

The Whitford Report discussed this issue at length. The preponderance of the submissions made to the Committee favoured registration over automatic protection, because this made the legal position clear by establishing priority and putting all concerned on notice of the rights claimed in the design. This view was strongly expressed by "many firms, trade associations and business and professional organisations representing the larger industrial organisations in general and the engineering industry in particular" (38). With the copyright system no-one in industry could put a product on the market without running the risk of litigation because someone somewhere had once made a drawing of which the product could be held to be a copy. Nor was it clear whether or not 'reverse engineering' and 'redesign' could be safely engaged upon, because if there was a design drawing there would be infringement, whereas if there was no drawing the same action would be lawful.

On the other hand, there had been submissions in favour of automatic rights as the better form of protection, in particular "from manufacturers or bodies representing industries producing products with eye appeal, for example the furniture and carpet manufacturing industries and the toy industry" (40). The delay between application and registration and the possibility of absence of protection before registration could be significant in certain industries. Further, "most small firms do not have specialist staff to concentrate on industrial property matters and in consequence find registration an administrative problem. Even when registration is sought there is a not insignificant question of official fees and agents' charges. This is one reason why small firms prefer automatic protection" (40).

Going against the tide of opinion expressed to it, the Committee opted for an automatic copyright-style system and the repeal of the registered designs legislation. This was partly because the Committee perceived declining use of the register of designs, from about 10,000 registrations *per annum* in the 1950s to less than 5,000 *per annum* in the mid-1970s. Repeal of the Registered Designs Act was a unanimous recommendation but there were various reservations with particular reference to industry's need for certainty as to questions of ownership and the age of the right. The Committee referred to its distinction of designs into Categories A and B. The majority of the Committee considered that damages for infringement should only be recoverable if the article in question had been marked so as to draw attention to the rights claimed. "The marking should be as in the Universal Copyright Convention, namely with the year date of first publication (marketing) and the name of the copyright proprietor at the time, placed in such manner and location as to give reasonable notice of the claim of copyright" (47). Linking the requirement of marking to the availability of remedies would, the Committee thought, avoid infringing Article 5(2) of the Berne Convention, which prohibits formal requirements for copyright to subsist in a work over and above its creation. Some members of the Committee were of opinion, however, that marking would not be sufficient for Category B designs and argued that for the functional design there should be a requirement of 'deposit'; that is to say, a registration process "in which formalities would be cut to the minimum but in which there would still be provision for public searching and preferably publication of deposits as well" (39). The general thrust of the Whitford recommendations was, nonetheless, in favour of automatic protection of all designs.

The 1981 Green Paper favoured protecting only aesthetic designs, but did not plump for one or other of either registration or automatic protection. Instead, attention was focused on the possibility of continuing dual protection. The Paper was clear that aesthetic designs should receive an automatic copyright protection, but pointed out that the registered designs legislation covered a slightly different field from copyright. Further, there had been "a certain resurgence of interest" (8) in the registered designs system, to judge from the number of applications for registration between 1978 and 1980 (over 5,000 each year). An important point was that if the EC sought a unified approach to design protection, it was likely to favour registration over automatic protection. A marking system, as proposed in the Whitford Report to meet the uncertainty of automatic protection, was, however, unacceptable. It was inconsistent with Article 5(2) of the Berne Convention, which prohibited the creation of formalities before the 'enjoyment and exercise' of copyright. The minority's 'deposit' system was also rejected: "[t]his would destroy the advantage of automatic protection, add to the burden and to costs and create practical difficulties for a number of trades" (7).

The 1983 Green Paper completed the move away from the Whitford position when it recommended that "there should be further examination of the feasibility, particularly cost, of replacing design copyright with registered designs" (23-4). The main arguments in favour of registration were the certainty which such a system created, and the absence of automatic protection in other countries. It was recognised that to extend registration had implications for the work load of the Patent Office, "given the large number of spare parts which are currently protected by design copyright", and consequently for public expenditure also. This objection might be met by the creation of a 'notification' system. The details of this were not spelt out in the Green Paper but presumably what was intended was a form of registration without full-scale examination of the design for novelty and the other features of registrability.

The issue of topography protection added further ingredients to cloud the question of whether or not registration should be the sole basis of protection. The US legislation of 1984 provided for a registration system but allowed a period of two years from first commercial exploitation of the topography before the right in an unregistered topography would lapse. The EC Directive permitted but did not require the adoption of a similar system in the Member States. It therefore left open the possibility of protection without registration at all.

The 1986 White Paper opted to continue a dual protection system. It noted that over 7,000 designs were now being registered per annum, suggesting that the register was of value to some sectors of British industry at least. A major disadvantage of extending the system as suggested in 1983, however, would be "the cost to users, particularly as in many cases a large number of registrations would be necessary to protect a single product" (19). Further there would be a call on the public purse which would not be justified "when even those areas of industry which would benefit from the protection obtained are far from unanimous in their support for the proposal and when the purpose of the proposal can be met in other ways" (19). The 1983 suggestion of a notification system was also rejected on the grounds of uncertainty and the possibility of meeting the problem in other ways. The White Paper also considered and rejected the possibility of an unfair copying law.

The White Paper accordingly took a step back to the Whitford recommendations by proposing a protection for designs which would arise without registration. This would co-exist with a system of registered designs but the two would be clearly distinguished in their fields of operation and in the nature of the protection offered. The uncertainty argument against automatic forms of protection "is not strong: original designs of all articles (except copyright artistic works) will be protected against copying and a competitor will know if he is copying" (22).

## Form Needed Before Protection Arises

This left the problem of the form in which the design required to be made before the protection arose. The Whitford Committee concluded that the position confining protection to drawings was "completely arbitrary and totally lacking in justification", and noted that "there was considerable support ... for the suggestion that copyright should subsist not only in original drawings but also in articles which start life in three dimensions, whether they are or are not works of sculpture or works of artistic craftsmanship" (42).

Accordingly it was recommended that designs should receive automatic protection in whatever form they were first created. Elsewhere in the Report it was recommended that the law should expressly accord copyright protection in the computer field, so covering the issue of designs produced by or with the assistance of a computer.

The 1981 Green Paper discussed these matters a little further and proposed that the copyright protection of industrial articles should be achieved by giving copyright to the articles themselves through "an expansion of the existing provision ... from works of artistic craftsmanship to works whose appearance is not solely dictated by the function they have to perform" (6). The copyright protection of drawings of functional articles would not extend to the making of three-dimensional items. There was no reference to this matter in the 1983 Green Paper, but the 1986 White Paper stated that the new unregistered design right would come into being with "(a) the first making of an article embodying the design; or (b) the first expression of the design in any independent form, such as a drawing or in a computer, from which an article embodying the design can be produced" (21). It would no longer be necessary for the designer to show that the design was a work of artistic copyright.

## Originality and Novelty

One other issue was raised sharply by the introduction of topography right, and that was the standard to be achieved by a design before it could claim protection. The registration system required novelty, while design copyright looked for originality. If there was to be a new right, as the 1986 White Paper proposed, which of these two different standards should be deployed? The White Paper talked of a standard of originality. Topography right introduced a third possibility, however. Under the US legislation on semiconductors, there was no protection for a topography which was not original or which consisted of designs that were "staple, commonplace, or familiar in the semiconductor industry" (s 902). The EC Directive set conditions that the topography be "the result of its creator's own intellectual effort and .. not commonplace in the semiconductor industry" (art 2(2)).

## For How Long Should Designs be Protected?

The Whitford Committee heard a variety of submissions on this point which it summarised as follows (41-2):

> There was a considerable body of opinion that whatever form of protection is provided in the industrial field it should not extend longer than the period provided for patents, which is currently 16 years but is to be extended to 20 years. There were suggestions, for example from the Chartered Institute of Patent Agents, that design protection should last only for five to ten years. This feeling was by no means universal however and there were other suggestions that a term of life plus 50 years is proper, at least in some fields. At the decorative end of the industrial

field a period of life plus 50 years was regarded as sensible, for there is no real reason for distinguishing such works from other artistic works. It was also argued that, as there is a tendency for designs to change over the years, a long term of protection could do no real harm. It was further suggested that any reference to the life of the designer in determining the term of protection is anomalous in the industrial field since many designs in that field are produced by employees, or under contract, and ownership is therefore normally in the hands of firms.

The final recommendation of the Committee was for a twenty-five-year term. The period would run from the marketing of articles bearing the design. A minority view was that the period for Category B designs should be shorter – between fifteen and twenty-five years. The upper limit of twenty-five years seems to have been derived from Article 7(4) of the Paris text of the Berne Convention, which provided that the period of protection for 'works of applied art' should be twenty-five years. The Committee also dealt with the interface between these terms and the full copyright term of artistic works: "It is only if, and when, articles to which the design has been applied are mass-produced and marketed that a shorter term for those articles should start to run" (51).

The 1981 Green Paper accepted that there was no justification for a full-length copyright term of protection, and proposed, following Whitford, that twenty-five years was appropriate, given its intention of protecting only aesthetic designs. Adoption of the twenty-five year term would "remove an obstacle to United Kingdom ratification of the latest Act (Paris 1971) of the Berne Convention" (6). Again the 1983 Green Paper did not consider this issue but seemed to assume that a fifteen-year period would apply.

A reduction in the term of protection was the major recommendation of the Monopolies Commission in its report on Ford replacement body parts. In that context, the Commission argued, even the fifteen-year protection for registered and registrable designs was too long (44):

> A monopoly lasting for 15 years is a severe limitation on competition. Demand for panels of a particular design continues only while cars of the model to which that design belongs are on the road. Models which are still in production 15 years after their introduction represent only a small proportion of cars.

*A fortiori* with the full-length copyright protection for unregistrable designs. The Commission accordingly recommended that the maximum period of protection for the designs of body panels should be five years; thus "the public interest in competition, Ford's interest in obtaining an adequate return on its investment and the public interest in allowing a reward for innovation and investment could all be met" (44). The Commission was careful, however, to limit its recommendations to the protection of body panels, and recognised that "consideration of the general reform of the law and copyright might reveal difficulties in the way of the particular amendments which we have recommended" (46).

The 1986 White Paper sought to resolve this question by returning to the view advanced by a minority in the Whitford Report, applying different periods of protection. But instead of having the different periods depend on the nature of the design, the White Paper proposed a distinction depending on whether or not the design was registered. Registered designs protection should be extended to a maximum of twenty-five years. This, of course, would be a protection only for designs with eye appeal. Unregistered designs, functional or aesthetic, would be protected for ten years from first marketing. The White Paper did not deal with the interface between the new right and copyright in the design, although it was obviously implied that copyright should generally not apply.

A factor which had clearly influenced Government thinking on this question was the new topography right. The US Act and the EC Directive set much shorter periods of protection than the copyright term: namely, in the US Act ten years from the date of registration or first commercial exploitation, whichever was shorter (s 904); and in the Directive, fifteen years from first fixation or encoding if there was no commercial exploitation in that period, or ten years from the first commercial exploitation or filing of an application for registration (art 7). Given that whatever replaced design copyright in the United Kingdom would probably be the means of protecting topographies, the US and EC legislation provided a formula for determining the vexed issue of term across the whole field as well as for semiconductor chips.

## Compulsory Licences

The general thrust of the Whitford proposals in 1977 was to confirm the legal protection of functional designs, while removing some of the objectionable features of the protection offered by copyright. As already noted, the Committee had received submissions on the anti-competitive effects of such protection, especially in relation to spare parts markets. While the Report refused to accept that these effects were reasons for allowing free use of functional designs, it did contain proposals to meet any abuse of monopolies which might occur. There had been submissions that there should be provision for compulsory licences "if the demands of the market are not met" (42), paralleling the position under the patents and registered design legislation. The Report concluded (50):

> [I]n the case of Category B designs, some form of compulsory licence should be available. It should be open to a person who wishes to copy the design (eg in order to supply replacement parts) to apply to the Comptroller-General of the Patent Office for a compulsory licence to do so on payment of the determined royalty on the ground that the United Kingdom market is not being adequately supplied by manufacture within the Community.

The point was not discussed in either of the Green Papers but was picked up in an altered form by the White Paper, which proposed that licences to manufacture the protected article would be available *as of right* during the final five years of the ten-year term of the new design right. "Furthermore, to avoid the risk that the five year absolute right might be abused, the Secretary of State will be given the discretionary power to order that the new right should become subject to licences as of right at any time if the Monopolies Commission should find that it has been exercised contrary to the public interest" (20).

This scheme considerably beefed up the Whitford proposal. A competitor became entitled to a licence, not because market demand was unmet, but simply because the design originator had enjoyed exclusive rights for five years, or because the exclusive right had been abused in a much broader way than simply failure to ensure adequate supply. So, whereas the Whitford Committee had only been moderately concerned with the anti-competitive effects of protecting designs, the White Paper scheme set up a situation where competition would take place if a competitor thought the market worth entering. As the White Paper put it (20):

> Competition from other manufacturers will however be possible reasonably early in the life of the product which is of particular importance for spare parts. It should be noted that, since the spare parts market only develops some time after the original product is first marketed, the effective period of protection for spare parts will be less than for other products.

The Whitford Report had also recommended that there should be provision for Crown rights in designs. This was the position under the Registered Designs Act, which the Whitford Committee wished to see repealed; but it also felt that "patents [in which there are Crown rights] and designs should on this point remain in step" (51). The Report did however doubt "whether it is necessary or intrinsically desirable for the Crown to have these powers for purposes other than defence" (50). The 1986 White Paper again took up the Whitford proposal in respect of the new unregistered design right, but made no concession in respect of the doubt stated in the Report.

## Scope of Protection: Monopoly or Copying?

The registration of a design confers a monopoly protection in that the owner may stop the marketing of any article made to the design without authority, even if there is no copying involved. Before there can be an infringement of a copyright there must have been copying. The Whitford Report argued that only protection against copying was needed. The matter was not discussed in the Green Papers but the White Paper adopted copyright language – 'reproducing the design' – in defining what would constitute infringement of the new design right. But no change was proposed for registered designs.

Topography right again raised interesting issues here. Under both the US and the EC legislation, the right was clearly to stop reproduction rather than a complete monopoly. This copyright-style protection was further limited by provisions that 'reverse engineering' – defined as reproduction for purposes of teaching, analysing or evaluating the concepts or techniques embodied in the topography, and then incorporating the results in a new topography – was not infringement (1984 Act, s 906; 1987 Directive, art 5(3)).

## Section 9(8) of the 1956 Act

The Whitford Report recommended the repeal of section 9(8) of the 1956 Act as difficult to apply and a cause of uncertainty. There was no dissent from this view in subsequent papers and reports.

## Foreign Law

Comparison of the British system of design protection with those found in other countries played a not insignificant part in the discussion of reform. One obvious reason for this is the international structure of intellectual property law generally, which, through the techniques of national treatment and reciprocity, has helped to establish a broad identity of rules around the world, enabling right-owners to enforce their rights everywhere. Therefore the law of any given state should not depart too far from the orthodox line, because it may disadvantage its own citizens abroad or give other countries' citizens advantages here which they would not enjoy at home. In a commercial sphere like industrial design, this has crucial implications for the domestic economy.

Closely related to this was the question of the EC position. National intellectual property rights can be a barrier to inter-state trade and, if commercial entities have different rights in the various parts of the Community, then competition may be artificially distorted. The Whitford Report observed that the Commission was examining the laws of the Member States with a view toward eventual harmonisation; it was therefore particularly important to look at the position in these countries.

The Report pointed out that the several potentially relevant international conventions in fact gave their member countries a fairly free hand in respect of industrial design protection. As a result, there was no overall consensus in the legal systems of the world. The Paris Convention only provided that "industrial designs shall be protected in all the countries of the Union" (Article 5 *quinquies*), and allowed foreign access to national registration systems, with an application in one Union country giving a priority in the others. The Hague Agreement concerning the International Deposit of Industrial Designs (1925, revised 1960) simply provided a means whereby a deposit in Geneva gave rise to protection in the member countries, and contained no substantive provisions on the content of design protection. The Berne Convention referred to works of applied art (probably a broader concept than the British 'works of artistic craftsmanship') and to industrial designs and models, but merely gave its members freedom to determine the application of their copyright laws to such works and to decide the term of protection for applied art, subject to a minimum of twenty-five years from the making of the work.

Whitford's survey of other legal systems suggested that most countries employing a registration system did not allow registration of functional designs (although the understanding of functionality varied), and refused automatic copyright to designs not in any way artistic. But in most European countries designs which achieved a high level of artistic worth could claim copyright. An even more detailed comparative study by Christine Fellner published in 1985 confirmed the overall conclusion, although she drew attention to some interesting developments in Australia and South Africa, where copyright protection along British lines was emerging. In the former country, however, the registration system had been amended to permit the protection of functional designs as well as those with eye appeal. Fellner also pointed to forms of protection other than registration and copyright which could be found in other countries, notably unfair competition laws, prohibiting 'slavish imitation' by competitors, and systems of 'petty or design patents' or 'utility models'. Under the latter functional products might be registered or deposited in a government office without a search for novelty or the conferral of extensive rights. In 1988, both South Africa and Canada legislated to remove copyright from industrial designs, subject to some limited exceptions, and Australia was preparing to do so. In these countries the chief focus for protection was to be the registration system, which in South Africa and Canada was very similar to the British model. A recognition that copyright went too far lay behind such reforms.

In general, therefore, as the Government prepared its Bill for submission to Parliament's scrutiny in 1987, the evidence from abroad seemed to favour protection of non-functional design through registration rather than automatic copyright.

# Chapter Four

## The Copyright Designs and Patents Act 1988

In bringing forward the Copyright Designs and Patents Bill in parliamentary session 1987-88, the basic policy of the Government was to provide a legal framework which would encourage innovation and competition. In design rights the encouragement of innovation meant, first, creating a legal structure in which investment in design would be rewarded by the grant of protection from unauthorised use of the design by others, and so give rise to advantage over competitors. Without protection there is less incentive to innovate; it is not cost-effective to invest in research and development of ideas if the end result immediately becomes common property. The existence of protection also forces competitors to engage in innovation if they wish to compete lawfully. Competition is fostered, however, by ensuring that this protection is not too strong; for example, in terms of the length of time for which it lasts. This care not to give over-powerful protection also provides a second level of encouragement for innovation, in that the owner of a right must continue to develop design ideas in order to retain or regain such competitive advantage as he may have gained by the initial design. There is a complex interaction between the protection and encouragement of innovation and competition, and a balance has to be struck if both are to be achieved. The Bill was clearly drafted with this balance in mind and, despite much criticism during its progress through Parliament, the Government's original proposals came through largely intact.

A further general point which informed the Government's whole approach in the Bill was the way in which the various parts of intellectual property law should interact in this area. Inventive products should be protected through the patent system. Products whose designs had aesthetic qualities should receive their protection from the registered designs system. Products without either of these qualities should not receive a greater but rather a lesser protection from the law. This point was constantly reiterated in answer to the parliamentary debates by the Government Ministers handling the Bill, Lord Beaverbrook and Mr John Butcher.

It was also clear that the Government had framed its scheme, first by generalising from the topography right first invented in the USA, and second, so far as concerned competition, by working from the problem of car spare parts as manifested by the British Leyland and Ford cases. The work of the Monopolies Commission and the European Commission on Ford, as well as the decision of the House of Lords in *British Leyland v Armstrong Patents*, had made apparent the need to remove impediments to a competitive market. As will be seen later in this chapter, the new law certainly removes protection from spare parts for cars. Accordingly, in this area the Government committed itself fully to the encouragement of competition rather than to finding a balance of interests. But this has not been achieved by legislating specifically for car spare parts but instead by the use of general words capable of covering many other situations. For a whole host of industries – the aerospace, electrical goods and motor manufacturing ones were most frequently mentioned in the parliamentary debates – the Bill therefore seemed to open up the prospect of what they regarded as piracy of their ideas and investment in research and development without any possibility of control or redress.

The critics of the Government scheme recognised the validity of the basic policy aims, while challenging at various points the means of achieving them and the balance sought to be struck between them. Some of the criticisms will be considered in detail below. The fundamental problem for the critics was their inability to achieve a consensus as to alternatives to the Government scheme, and this was a rock upon which attacks on the Bill frequently foundered.

The 1988 Act deals with the problem of design protection on three fronts: first, by severely limiting the ways in which the copyright subsisting in designs may be infringed; second, by amending the Registered Designs Act to make registration more attractive to users; and third, by creating a new unregistered design right of much more limited effect and duration than either copyright or registration. The basic rules of the 1956 Act which led to *British Leyland v Armstrong* are, for the most part, still to be found in the 1988 Act. The belief was that as a result of new provisions in the 1988 Act artistic copyright would cease to be of major significance for industrial designs, although it has not been removed altogether. Copyright subsists in original artistic works. Artistic work includes what are now termed graphic works, irrespective of artistic quality and a graphic work includes any drawing, diagram, map, chart or plan. In relation to an artistic work the restricted act of copying includes (i) reproduction in any material form, which can be in any way and for any purpose [again, still contrast the limited scope of infringement of a registered design]; and (ii) the making of a copy in three dimensions of a two-dimensional work. However, the non-expert test as a defence to this kind of infringement was repealed. The White Paper had said that the spare parts exception propounded by the House of Lords in *British Leyland* (i.e. no derogation from grant) would also be removed. This was not done in so many words and indeed section 171(3) provides that nothing in the Act affects any rule of law preventing or restricting the enforcement of copyright, on grounds of public interest or otherwise. Given the legal doctrine of precedent, it is therefore still possible to plead no derogation from grant against copyright enforcement. The hope was that, at any rate with regard to the problem of spare parts, it would not be necessary to do so, because other provisions of the Act took care of the matter. In fact, the defence has been used since the entry into force of the 1988 Act, but not with any particular success and certainly not in any case about spare parts (MacQueen *et al* 1993: para 1133).

# THE COPYRIGHT DESIGNS AND PATENTS ACT 1988

## Restricting Copyright

This is done principally through sections 51-52 of the Act. Section 51 says what is *not* infringement of copyright in a design document or model recording or embodying a design for anything other than an artistic work or a typeface. A design document is defined as any record of a design, whether in the form of a drawing, a written description, a photograph, data stored in a computer or otherwise. The *ejusdem generis* principle of statutory interpretation (Bennion 1984: 828) suggests that the 'otherwise' covers only two-dimensional design records like the others actually mentioned in the definition. Three-dimensional *Hensher*-style prototypes, insofar as they have copyright at all, are covered by the earlier words in the section referring to 'models' embodying a design. There remains the question of what constitutes a design: it is defined in the Act as the design of any aspect of the shape or configuration (whether internal or external) of the whole or part of an article, other than surface decoration.

It is important to grasp that section 51 recognises copyright in design documents. But the significance of this is carefully restricted. It is not infringement of the copyright to make an article to the design or to copy an article made to the design. This has the effect, broadly, that the unauthorised making of three-dimensional articles by copying either directly or indirectly from a design document cannot give rise to a copyright action.

If there is a design document for an artistic work – for example, an architect's plans for a building or a design for a work of artistic craftsmanship such as certain types of clothing, furniture, sculpture, or photographs – or for a typeface, then under section 51 that design enjoys the full protection of artistic copyright. The protection will last for the lifetime of the author plus the post mortem period of protection. The author and his successors will be able to take action throughout that period to stop reproduction of all kinds, including the making of three-dimensional copies. Artistic copyright also covers surface decoration applied to an industrial product.

It will therefore be vital to distinguish when a drawing or model is the design of an artistic work and when it is simply the design of an article. Only in the latter case will the full protection of copyright be unavailable. The most crucial things to keep in mind in approaching the question are (i) the distinction between shape and configuration on the one hand and surface decoration on the other; and (ii) which three-dimensional works attract copyright (i.e. works of architecture and artistic craftsmanship, sculptures and casts and moulds made for purposes of sculpture). Questions arose even before the 1988 Act as to whether commercial products may have copyright in themselves as such works. It is recognised that certain types of furniture, clothing and toys can claim the status of artistic works, but in general such questions have been answered in the negative by the courts where the product has been intended for a mass market, as was the case in *Hensher* and *Merlet v Mothercare* [1986] RPC 115. Prototypes of plastic dental impression trays were refused copyright as either sculptures or models by Whitford J in *J & S Davis (Holdings) Ltd v Wright Health Group* [1988] RPC 403. The New Zealand case of *Wham-O Manufacturing Co v Lincoln Industries* [1985] RPC 127 does provide a contrary instance, holding that the Frisbee was an artistic work as an engraving and that wooden mould prototypes for the toy were sculptures. Given the limitations of the new design right, we may see renewed efforts by manufacturers to bring their products within the scope of three-dimensional artistic copyright. Consider this paean of praise for the modern car emanating from Mr. Richard Page MP during the parliamentary debates (24 May 1988, HC Standing Committee E, cols 238-9):

> I trained as a mechanical engineer and to me an exhaust system is a thing of beauty. It has an artistic ability of its own, it is not simply a functional piece of apparatus... Have hon Members seen how, in the latest racing cars, the exhaust systems blend to make a harmonious whole? They may not appreciate them, but to a mechanical engineer they are things of beauty and artistic design... How can anyone not admit that a car body shell is a thing of beauty? Design teams spend fortunes on producing harmonious shapes and ensuring that they are pleasing and artistic and have popular appeal. It always seems that something that is useful is not artistic, but that if it cannot be used it is artistic. Sir William Lyons, the creator of Jaguar cars – who sadly has now passed away – always had delivered to his home the latest sculpted model; he put it in the garden, spent hours looking at it and then sent it back for the shape to be altered. Only after hours and hours of studying it could he feel that he had something that represented the peak of artistic ability and that would appeal to the public. It is incredible to me that a sculpted body shell is not regarded as a thing of beauty. In Italy Ferina, Ital and Bertona have design studios that are trying to create body shells which are sufficiently attractive to be purchased by the manufacturer and then by the public.

If such arguments succeed in convincing a court that a car body or exhaust pipe is a sculpture or a work of artistic craftsmanship, the designer will then run into the obstacle of section 52, which deals with artistic works which are exploited by being made into articles by an industrial process (i.e., making fifty or more copies) and marketed (exposed for sale or hire). Once industrial exploitation has occurred, the copyright owner has his protection restricted; twenty-five years later, a term presumably derived from the Berne Convention provisions on works of applied art, he loses the right to claim that the unauthorised making of similar articles infringes the copyright. So, while copyright might still be a means of protecting industrial products founded on an artistic work, there is a restriction on the duration of the protection paralleling the maximum for a registered design under the 1949 Act as amended. The aesthetic element of the original work justifies the equivalence.

The section also meets the *Popeye* situation where a design starts out as an artistic work pure and simple, and only later becomes a design document for manufacturing articles by an industrial process. Modern character merchandising in respect of children's entertainment may also pose some interesting tests for the application of section 51. It is often the case that a character is presented in a co-ordinated campaign across a variety of media – a TV cartoon or puppet series, a comic or magazine centring on the character or characters, models of the character and other merchandise using representations of the character as surface decoration (Adams 1987). Here it may be difficult to avoid the conclusion that, if there are initial drawings of the characters, they are design documents since they are to be exploited by the production of articles made to the design. But are they also designs for artistic works, in that they will provide the basis for the drawings to be used in cartoons, comics and the surface decoration of other articles, all of which will have their own copyright? Determining the length and scope of protection in cases like this will be a complex matter, to say the least.

Section 52(4) empowers the Secretary of State to make orders for excluding from the operation of the section articles of a primarily literary or artistic character. Examples of what these orders cover are the production of books and the exploitation of a painting or a photograph by the manufacture of posters, books and posters both being articles. It is not the legislative intention that the author should lose his full copyright by this kind of commercial exploitation, which can be seen as connected with the essence of copyright rather than industrial design law.

The intended effect of sections 51 and 52 is that normally an industrial design cannot be protected from unauthorised industrial exploitation through copyright. While it is possible to figure some difficulties, it seems likely that this will cover the kind of case which was causing the major problems before the passage of the 1988 Act. In such situations protection against unauthorised similar products must now be sought through registration under the Registered Designs Act 1949 as amended, or through the new unregistered design right.

## Amendments to the Registered Designs Act

There were a considerable number of amendments to the Registered Designs Act 1949, many of which were intended to align the protection of registered and unregistered designs. These will be examined in the discussion below of the new right. There are, however, some changes not of this character. Most importantly, the maximum period of protection was increased from fifteen to twenty-five years, in five five-year blocks, renewal being necessary at the end

TABLE 5.1 APPLICATIONS AND REGISTRATIONS UNDER THE REGISTERED DESIGNS ACT 1949 1984–1993

| Year | Applications | Registrations |
|------|--------------|---------------|
| 1984 | 7237 | 6697 |
| 1985 | 7395 | 6546 |
| 1986 | 7844 | 7167 |
| 1987 | 8646 | 7140 |
| 1988 | 8748 | 8049 |
| 1989 | 9317 | 8945 |
| 1990 | 8566 | 9171 |
| 1991 | 8074 | 6271 |
| 1992 | 8267 | 8175 |
| 1993 | 8179 | 8301 |

(*Source: Annual Report of the Patent Office 1993-94* (HC 618): 18)

of each period if the registration is not to lapse. The aim was to make registration a more attractive option as a stronger form of protection. Table 5.1 suggests that in this regard the Act has been a success, inasmuch as the trend of increasing numbers of applications and registrations noted in the 1986 White Paper (above, 56) accelerated immediately after the passage of the legislation, although it now appears to have steadied at over 8,000 applications and 8,000 registrations *per annum*.

Some caution should be observed, however, before adopting this conclusion wholesale. In particular the suggestion that the longer term of protection is a factor in encouraging businesses to register their designs seems not to be borne out upon consideration of another set of figures (Table 5.2), which shows how many right-holders have extended their registrations into the second and third five-year periods of their registration. For obvious reasons, it is not yet possible to give figures for extensions into the fourth and fifth periods now available under the 1988 Act.

The Patent Office (1994: 23) estimates that these figures represent about 40% and 19% respectively of the number of applications filed five and ten years previously. This suggests that for many businesses five years' worth of exclusive rights in a design is enough, and that for a very substantial majority ten years is quite sufficient. What is harder to detect from the figures alone are the reasons why renewals do not occur: cost, commercial failure of the design or of the right-holder's business, and simple administrative inefficiency are possibilities as well as conscious decisions that the purpose of protection has been achieved. It is unlikely, however, that having the possibility of protection for twenty-five years is a significant factor in the decision whether or not to register a design in the first place.

To be registrable a design must still have eye appeal and not be functional. The *Amp* decision, that the eye in question is the eye of the customer, is reinforced by a further test for registration, namely, whether the article to which the design is to be applied is one where appearance is not material. Appearance is not material "if aesthetic considerations are not normally taken into account to a material extent by persons acquiring or using articles of that

TABLE 5.2 EXTENSIONS OF DESIGN PROTECTION UNDER THE REGISTERED DESIGNS ACT 1949 1984–1993

| Year | 2nd Period | 3rd Period |
|---|---|---|
| 1984 | 1751 | 672 |
| 1985 | 1947 | 722 |
| 1986 | 2354 | 732 |
| 1987 | 2374 | 734 |
| 1988 | 2706 | 968 |
| 1989 | 2991 | 984 |
| 1990 | 2892 | 968 |
| 1991 | 3079 | 1281 |
| 1992 | 3465 | 1196 |
| 1993 | 3480 | 1292 |

(*Source: Annual Report of the Patent Office 1993-94* (HC 618): 23)

description and would not be so taken into account if the design were to be applied to the article" (1949 Act s 1(3), added by 1988 Act s 265).

However, the most significant development in registered design law since the 1988 Act has not come about as a result of these legislative adjustments. In *Ford Motor Co Ltd and Iveco Fiat SpA's Applications* [1993] RPC 399, affirmed sub nom *Ford Motor Co Ltd's Design Application* [1994] RPC 545 (DC), affirmed sub nom *R v Registered Designs Appeal Tribunal ex parte Ford Motor Co Ltd* [1995] 1 WLR 18 (HL) the interpretation of the word 'article' and its application to spare and accessory parts for motor vehicles came under consideration. The question was whether the designs of spare parts for motor vehicles had been "applied to articles" as required by the 1949 Act. It was finally held by the House of Lords, confirming the views taken in the courts below, that in general such parts had no reality as articles of commerce apart from the vehicle itself and so, not being articles to which a design had been applied, there could be no registration of the design. The judgment turns on the words used in the 1949 Act to define "article" ("any article of manufacture [including] any part of an article if that is made and sold separately" – s 46(1)). But applications in respect of the designs of a wing mirror, a vehicle seat, a steering wheel, wheels and wheel covers were allowed to proceed, these being features where substitutions of a character distinct from the part being replaced were possible, while leaving the general shape and appearance of the vehicle itself unaffected. The House held that there was an essential difference between an item designed for incorporation in a larger article, whether as an original component or a spare part, which would be unregistrable as an article, and an item designed for general use, albeit aimed principally at use with the manufacturer's own artefacts, which would be registrable as an article.

With this understanding of the scope of the word "article", the arguments about the scope of the protection of spare and accessory parts under registered design law cease to be of great importance. Many, if not most, spare parts are not registrable at all, regardless of any other consideration which might otherwise apply. The decision is therefore a remarkable one in many ways. The designs of motor vehicles and their parts have been amongst the classes of article forming the subject matter of the largest number of filings in the Designs Registry (Patent Office 1994: 18); so there seems certain to be an appreciable adverse effect on the number of registrations in future. On a simple reading of the statutory definition, and as a matter of fact, it seems impossible to say that replacement parts are not "made and sold separately"; as this book has demonstrated, there is a thriving marketplace in such items. The House of Lords was concerned, however, that such a simple reading would lead to any part of a larger whole being potentially an article with a registrable design, and glossed the statute to say that, before such a part could be an article, it had to be in effect intended by the designer that it be made and sold separately. A car was not designed in parts but as a whole. Again, however, this seems not wholly in tune with commercial reality, where motor vehicles are indeed assemblies of parts and the replacement of those parts over the life of the vehicle is anticipated and intended to be met by the manufacturer. The judgment thus stands alongside the *British Leyland* case as a reaction against the intellectual property protection of spare parts, and gives rise to a curious, wholly arbitrary and uncertain division between those parts which are articles in law and those which are not; all on a basis which would never occur to any sensible businessman or designer. (For detailed criticism of the judgments of the courts below on this point, which were affirmed by the House of Lords, see Laddie, Prescott & Vitoria 1995: paras 30.18-30.26.)

## Unregistered Design Right

A further objection to the *Ford* decision is its failure to consider the connection with the new form of protection introduced by the 1988 Act, unregistered design right. It is defined as a property right subsisting in an original design. Design means the design of any aspect of the shape or configuration (whether internal or external but excluding surface decoration) of the whole or part of an article (aligning with the definition in section 51). There is no statutory definition of the meaning of the word 'article' in relation to unregistered design right, but it should be noted that the design protected may be of *part* of an article. It is therefore interesting to speculate as to the effect of the *Ford* decision in this area. It would be curious if the word 'article' ended up with different meanings in two contexts so closely-related, but otherwise the elaborate provisions in the Act making exceptions from protection for spare parts, on which so much energy was expended in 1987-8 (see below, 73-8), may be near to redundancy.

Turning to the other aspects of the definition of unregistered design right, 'shape or configuration' is a concept already familiar from registered design law as referring to the three-dimensional aspects of the product, and distinct from two-dimensional pattern or ornamentation which is applied to the surface of the article. The statute also expressly excludes surface decoration from design right as part of this emphasis on the three-dimensional. Methods or principles of construction are excluded from design right as well, being protectable by patent if at all. Design right therefore concentrates on the appearance of the product rather than on the way it is made. But there is no requirement that the appearance of the product should appeal to the eye, either of the customer or of anyone else. Indeed, it is clear from the section that the design feature need not be visible in the finished article. Functional designs therefore attract design right, although it is not confined to such designs.

The emphasis on the three-dimensional as the subject-matter of unregistered design right leaves at least one problem area, namely electronic circuit designs. John Reynolds and Peter Brownlow (1994: 400) have commented as follows:

> An electronic circuit diagram does not specify the 'shape', only the constituent components and their relative connections – what route the connections take and the positioning of the components is generally not recorded in the schematic Accordingly, unless 'configuration' is to be interpreted as meaning something more akin to constitution, structure or composition, departing perhaps from the established design law, the schematic of an electronic circuit diagram may well escape the definition of a design document.

They observe that a similar problem may arise with mechanical engineering designs such as pneumatic and hydraulic circuits and with chemical or process flow diagrams. Should such designs fall outwith the scope of unregistered design right, then they may still be protectable through copyright, although that will still give rise to the issue of how that right may be infringed by three-dimensional reproductions (see above, 37-8). Design right does apply to semiconductor topographies, although these are essentially patterns fixed or etched upon semiconductor material, rather than shapes or configurations. The availability of the protection is however the result of express legislation, the Design Right (Semiconductor Topographies) Regulations 1989, which apply the design right provisions of the 1988 Act to this subject matter.

Another interesting possible subject of unregistered design right has been suggested by Sean Hird and Michael Peeters (1991), namely synthesised molecules of DNA, the products of genetic engineering. The interest of this lies in the difficulty which biotechnology products have faced in obtaining intellectual property protection, in particular through patents. Hird and Peeters (1991: 336-7) argue that "each of the four classes of nucleotide which make up a DNA molecule has a unique shape which can be distinguished under an electron microscope. Therefore a recombinant DNA molecule will have a unique shape insofar as it consists of a unique sequence of nucleotides". Since there is no requirement that a design be visible, there seems no reason why a molecule should not be an article, and that accordingly the genetic engineer may claim unregistered design right. The claim might however fall foul of the exclusion of methods or principles of construction from design right, which may be applicable to a nucleotide sequence.

## Original and not Commonplace

Designs must be 'original' to attract design right. But it is also provided that "a design is not 'original' for the purposes of [unregistered design right] if it is commonplace in the design field in question at the time of its creation" (s 213(4)). This compromise between the 'novelty' required for design registration and the copyright standard of originality clearly came from the topography right formulation first expressed in the US Semiconductor Chip Protection Act 1984. The first case on unregistered design right, *C & H Engineering Ltd v F Klucznik & Sons Ltd* [1992] FSR 421, confronted the meaning of the new standard. The subject matter for discussion was pig fenders, which are devices to stop piglets leaving the sty while enabling the sow to step over into the field outside the sty. It is important that the fender be shaped so that the sow's teats are not scratched as she steps over it. The fender in the case solved this problem by having a two-inch rounded metal tube placed around its top edge. Aldous J noted: "By 1990 pig fenders were commonplace and had been made in metal and wood. In essence [the farmer] wanted a commonplace pig fender with a metal roll bar on the top ... the only part of the pig fender shown in the drawing which was not commonplace was the 2 inch tube on the top" ([1992] FSR at 428). Given these observations, it is puzzling that he nonetheless went on to say that originality in design right is the same as in copyright. (See the same point also made in Tootal 1990: 187-8; Laddie Prescott and Vitoria 1995: para 40.12.) The test meant that the design must not be copied from another design but be the independent work of the designer ([1992] FSR at 427). Aldous J then referred to the provision about commonplace designs and concluded that "for the design to be original it must be the work of the creator *and* that work must result in a design which is not commonplace in the relevant field" ([1992] FSR at 428, emphasis supplied). This is correct and, it is submitted, means that the test of originality in unregistered design right is *not* the same as in copyright. The exclusionary definition in section 213(4) means that a design which is 'commonplace' in the design field in question at the time of its creation is not original and so not protected. Thus the issue concerned, not the whole of the pig fender, but just the roll bar. Under section 213 design right subsists in designs, and a design may be the design of the whole *or part* of an article. Design right could therefore have subsisted in the roll bar alone. It is doubtful whether the rest of the pig fender had any design right.

This conclusion about the restricted meaning of originality in relation to unregistered design rights of course raises difficulties. An interesting example of the type of problem likely to be debated before the courts can be found in the *Interlego* case, where Lord Oliver noted of the Lego brick that "inevitably a designer who sets out to make a model brick is going to end up by producing a design, in essence brick shaped" ([1989] AC at 246). Two points arise: one, is there a design field in toy bricks, the other, is the brick shape a commonplace one? The test may be comparable to the novelty test for registration of a design. It is certainly more restrictive than the copyright originality concept, where banality and the rehashing of old ideas have never been held unoriginal so long as an independent skill, labour and mode of expression was used. It seems possible, therefore, that commonplace features of a design will be protected by copyright, subject of course to the limitations as to what constitutes infringement thereof under section 51. Where industrial copying is being challenged by means of design right we can therefore expect some interesting disputes on whether or not the first design is commonplace.

The requirements that a design be original and not commonplace may be of importance for the possible protection through unregistered design right of recombinant DNA molecules since, as Hird and Peeters (1991: 338) observe, "DNA molecules can only ever be comprised of four classes of molecules". They suggest that the "high level of technical skill necessary to create a DNA molecule with a unique nucleotide sequence" should satisfy the criterion of originality; and presumably, if the sequence is unique, the 'not commonplace' test as well.

Setting aside the 'not commonplace' aspect, other issues about originality which arose in the *Interlego* case may still cause difficulties for unregistered design rights. The drawings in the case were based on drawings of previous models of the Lego brick. Interlego argued that, despite their derivative nature, the drawings were original for copyright purposes because their making had involved skill and labour. The Privy Council held that skill and labour were

# THE COPYRIGHT DESIGNS AND PATENTS ACT 1988

not sufficient for originality but that there was a need for some material alteration or embellishment in relation to the first work. The presence on the drawings of explanatory figures and words did not confer originality upon them, if there was no significant difference in the visual impression created by the two works. Given that unregistered design right is primarily concerned with the appearance of products, is it still the case that there must be visual originality in the second product? Further, what happens in the case where a design commences life as a prototype and drawings are subsequently made from that? Is this sufficient alteration? Lord Oliver made some comment on this ([1989] AC at 263):

> In this connection some reliance was placed on a passage from the judgment of Whitford J in *L B Plastics Ltd v Swish Products Ltd* [1979] RPC 551 at 568-569 where he expressed the opinion that a drawing of a three dimensional prototype, not itself produced from the drawing and not being a work of artistic craftsmanship, would qualify as an original work. That may well be right, for there is no more reason for denying originality to the depiction of a three dimensional prototype than there is for denying originality to the depiction in two dimensional form of any other physical object.

On the other hand, what about the design which is worked out over time through a series of drawings or other representations? Will the ultimate design in the series be sufficiently different to be original under the *Interlego* test?

## 'Must Fit and Must Match' Exceptions

There are two other features of designs which may be excluded from the protection of design right – the 'must fit' and 'must match' exceptions, concerning which there was considerable debate as the measure passed through Parliament. The design of features of shape or configuration of an article which are there to enable the article to fit with, or match, the appearance of another article is excluded from unregistered design protection. 'Fit' is an encapsulation of the statutory formula, which is "features of shape or configuration of an article which enable the article to be connected to or placed in, around or against another article so that either article may perform its function" (1988 Act s 213(3)(b)(i)).

'Match' summarises "features of shape or configuration of an article which are dependent upon the appearance of another article of which the article is intended by the designer to form an integral part" (1988 Act s 213(3)(b)(ii)). The exceptions also extend to the scope of the protection for a registered design. 'Must match' is dealt with explicitly in an amendment to the Registered Design Act 1949 (s 1(1)(b)(ii)), while 'must fit' apparently falls within the existing exclusion from registration of designs dictated solely by function (*Ford Motor Co Ltd and Iveco Fiat SpA's Applications* [1993] RPC 399, affirmed sub nom *Ford Motor Co Ltd's Design Application* [1994] RPC 545).

The aim of these exceptions, as the Government made clear in Parliament, was to tackle the problem of spare parts for motor cars which had been at the heart of the *British Leyland* case and the Monopolies Commission report on Ford. Despite valiant and repeated efforts in both Lords and Commons, the legislation was not confined to that specific situation and, as a result, the precise scope of its application is far from clear. Plainly it covers the *British Leyland* case: an exhaust pipe is an excellent example of a design with features of shape and configuration which are there to enable it to fit another article (i.e. a car). The decisions of the lower courts and tribunals in *Ford Motor Co Ltd and Iveco Fiat SpA's Applications* [1993] RPC 399, affirmed sub nom *Ford Motor Co Ltd's Design Application* [1994] RPC 545 (DC) also confirmed that motor vehicle body panels could not be registered as a result of the 'must match' exception.

What then is the scope of the exceptions outside the motor industry? 'Must fit' applies to topographies, and so the pattern of the interfacing area of a semiconductor is unprotected. The *Financial Times* reported on 22 February 1995 that the New World Group was suing Creda for infringement of unregistered design rights in double ovens to be built into waist-high kitchen units; presumably at least part of the designs involved would be excluded from protection as 'must fit' features. An example which was much discussed at the time of the Bill was that of Lego bricks. In the *Interlego* case, Lord Oliver commented that "there is clearly scope in the instant case for the argument that what gives the Lego brick its individuality and the originality without which it would fail for want of novelty as a registrable design is the presence of features which serve only the functional purpose of enabling it to interlock effectively with the adjoining bricks above and below" ([1989] AC at 246). Equally clearly there is scope for arguing that these features fall precisely within the scope of the 'must fit' exception to unregistered design right. If we combine that with the point made earlier about the 'commonplace' features of the brick, it looks very much as though Lego bricks would be unable to claim design right in any respect. If this conclusion is correct, it would lead to a curious anomaly in that a design could be registrable yet unable to attract design right. But in the course of the parliamentary debates (24 May 1988, HC, Standing Committee E, col 242) the Government accepted that the *Interlego* case had been correctly decided on the registrability point and that this would be unaffected by the changes to the Registered Designs Act.

Nonetheless, by the conclusion of the Bill's progress through Parliament the 'must fit' exception seemed to have gained a grudging acceptance in many quarters otherwise critical of the provisions. Views may have been swayed by the Government's repeated argument that this exception did not deprive spare parts of design protection: only features which permitted the spare part to be fitted to the original piece of equipment were unprotected as a result. Mr Butcher provided the example of an agitator for a Hoover vacuum cleaner to illustrate the argument. The agitator was connected to the cleaner by fittings at either end and only these features were caught by the exception. The remainder of the agitator – some 80% of it – was protected by design right (*Parliamentary Debates*, 14 June 1988, HC Standing Committee E, col 532). Not all critics were appeased, particularly the manufacturers of large and complex machines which were essentially assemblies of components, each

## THE COPYRIGHT DESIGNS AND PATENTS ACT 1988

designed to fit into its place in the overall structure and all of whose design features could be said to be there to be placed in, around or against another article. In particular, the aerospace and motor manufacturing industries seem to have maintained a steadfast opposition on this basis to the 'must fit' exception.

The heaviest fire in the parliamentary debates was however reserved for the 'must match' exception. By contrast with 'must fit', the Government always maintained that 'must match' was indeed a blanket exclusion of protection for spare parts falling within its scope. Thus, for example, Lord Beaverbrook (29 March 1988, HL, col 699):

> The must match exception is intended to prevent monopolies arising in the first place, and to preserve the benefits of competition. Although design right is only a right to prevent copying, it is quite clear that in circumstances where a competitor has no choice but to copy if he is to produce a part which will match, then if there were no must match exception he would be completely shut out of the market. This is not a question of abuse but of basic policy. And I have to say that this Government does not wish to create monopolies in this way in any sector of industry... And we should be quite clear about this: the absence of a must match exception would enable competition in certain kinds of product to be totally frozen out. In our view that is not the way that the markets should operate.

This made it clear that where 'must match' applied there was no question of partial protection for the part in question. All of it was open to the copyist.

The attack on this provision was made on a number of fronts. First and foremost was the scale of research and development costs incurred before a product was put on the market. Figures of £600 million and £80 million were mentioned for General Motors' then-new Vauxhall Cavalier and the body-work of Ford's Sierra respectively (*Parliamentary Debates*, 14 June 1988, HC Standing Committee E, cols 522 and 543). How could this be recouped if a copyist could simply appropriate the design ideas produced through expenditure of such magnitude? Even research and development on a massive scale did not necessarily lead to products which were either patentable or had registrable designs. There was a need for some lesser protection. An innovative and massively expensive machine like the Concorde aircraft had only produced twelve patentable parts (*Parliamentary Debates*, 25 July 1988, HC, col 41). The copyist who did not have to engage on initial research and development was bound to be able to supply the parts more cheaply, and deny the original equipment manufacturer the opportunity to compete in the market which he had created. Further, the copyist was only interested in the 'fast-moving' parts for which there was a high demand from early in the life of the machine, and did not have to undertake the manufacture and stocking of the 'slow-moving' parts. The undercutting of the original manufacturer deprived him of the incentive to innovate in design. The total effect would include reduction of employment opportunities, especially given that many copyists operated from bases overseas. The copyist would not have the same incentive as the manufacturer seeking to protect his good name to ensure the quality and safety of his parts, and encouraging copying was therefore a threat to the consumer interest.

Replies to these arguments were as various as the arguments themselves. Research and development costs were recovered in the pricing and selling of the original equipment, not by participation in the after-market. The consumer stood to benefit from being able to obtain replacement parts more cheaply. As to quality and safety, consumer protection legislation existed to ensure this, and it was not the function of intellectual property law to protect the consumer. The argument against the independent producers of spare parts was undermined by the fact that the independents were often the originators from whom the original manufacturers purchased products for their own purposes, while engaging in the manufacture and supply of spare parts for other people's products themselves. The creation of a legal environment which encouraged rather than restricted that kind of activity could have a beneficial effect on the employment situation by opening up the prospects of new jobs.

The defence of 'must match' was by and large successful. Little of it was amended as a result of the parliamentary discussions. The principal adjustment was to ensure that the exception did not apply to 'families' of items such as dinner services and crockery sets. As originally drafted, it appeared that 'must match' would permit the manufacture and supply of replacement plates or crockery without the consent of the manufacturer of the set. The requirement that the replacement article should be an 'integral part' of the other article whose appearance it is to match is intended to prevent this result. On the other hand, both the 'must match' and the 'must fit' exceptions apply to kits of parts which when assembled form a complete article (a common example is an item of self-assemble furniture such as was the subject of *L B Plastics Ltd v Swish*).

Although the House of Lords in the *Ford* case found it unnecessary, as a result of the decision that spare parts were not articles, to consider the scope of the 'must fit' and 'must match' exceptions, the judgments below do offer some helpful discussion. The most important points to emerge are that 'must match' is not a blanket exclusion of all accessory parts, and that the requirement of dependency upon the appearance of another article of which the article in question is to be an integral part restricts the scope of the exception. Deputy Judge Jeffs in the Registered Designs Appeal Tribunal found the amended provisions of the 1949 Act on 'must match' "undoubtedly ambiguous and obscure" ([1993] RPC at 421). Like the Registrar at first instance, he therefore invoked the ruling of the House of Lords in *Pepper v Hart* [1993] AC 534 and found guidance in the parliamentary debates on the legislation ([1993] RPC at 413-414, 421-422). The debates made it quite clear that the intention of Parliament had been to deny design protection to car body panels. While such panels might be sold as independent items, they had necessarily to be the same in appearance as those which they replaced. The design of the panel was subordinate to the design of the vehicle as a whole. Some other items such as wheel covers, steering wheels, seats and wing mirrors were, however, ones where an owner of a vehicle might choose between alternatives, for example, to give a car a sportier appearance or increase its comfort. Thus the appearance of parts such as these was not integral to the appearance of the vehicle as a whole, and the designs fell outwith the scope of the 'must match' exception.

The Divisional Court rejected the need to invoke *Pepper v Hart*, McCowan LJ commenting that he did "not consider that in this case the ministerial statements clearly disclose the mischief aimed at or the legislative intention. In fact I have sympathy with Mr Silverleaf's [counsel for the Registrar] description of them as 'long and diffuse and varied'" ([1994] RPC at 554). It was nevertheless held, in agreement with Judge Jeffs, that in considering whether or not an article's features were dependent upon the appearance of another article, the whole of the latter article should be considered and not that article minus the part in issue.

These decisions appear to be a correct view of the intended effect of the exceptions. It is important to note that a number of motor vehicle accessories were thus held to fall outwith the exceptions, having on basically similar grounds also been accepted as independent articles of commerce. Such accessories would therefore also receive protection from unregistered design right if registration was for any other reason not used or achieved. It can accordingly be seen that the exceptions do not wholly remove such protection for spare parts for motor vehicles as may still survive the decision that in general such parts are anyway not articles.

## Functional Articles other than Spare Parts

The parliamentary debate on the 'must fit' and 'must match' provisions of the Bill concentrated on functional designs and spare parts, as the issues which had caused the greatest difficulty in practice, and the problems of which it was intended to solve. But the new design right also applies to free-standing items of all kinds (for example the pig fender and roll bar discussed in the *Klucznik* case), and covers aesthetic designs as well as functional ones. In this way it meets the problem first tackled legislatively in the Design Copyright Act 1968 – the protection of the unregistered design from industrial copying. Given that registration of a design will remain a relatively cumbersome and expensive process, this alternative protection is clearly necessary. It helps the manufacturers whom it was sought to protect under the 1968 Act, who are unlikely to be caught by any of the exceptions to design right. As will be seen below, the protection is not so extensive as under either the amended 1949 Act or the 1968 Act, particularly in respect of duration, and, accordingly, the manufacturer will have to make a choice whether or not to register his design. That choice will presumably depend on the resources of the designer and the expected market life of the product, and must be made before the production process begins. The manufacturer will as a matter of course have design right and, for what it is worth in the light of sections 51 and 52, copyright from the time that there is a design document, but if a product made to that design is marketed prior to registration, then the right to register will be lost.

## Form Needed Before Protection Arises

A question which unfortunately continues to be difficult under the new Act is whether design right can subsist in a prototype which is excluded from copyright by the *Hensher* case. Section 213(6) states that "design right does not subsist unless and until the design has been recorded in a design document or an article has been made to the design". Is a model an article made to the design? This seems doubtful, particularly where the design is actually being evolved by work on the model. Design document is defined as in section 51; that is, as "any record of a design, whether in the form of a drawing, a written description, a photograph, data stored in a computer or otherwise" (1988 Act s 263(1)). The 'otherwise' might cover the working model, if there can be escape from the *ejusdem generis* principle of statutory interpretation, under which general words like 'otherwise' are to be limited in their meaning by any more specific words which immediately precede them. Given that the other specific examples are two-dimensional, applying the interpretative principle makes it difficult to bring three-dimensional objects within the scope of the definition of a design document. Moreover, section 51 actually refers to a 'design document or model'. Since that section uses exactly the same definition of 'design document', it is arguable that the phrase cannot cover 'model'. Accordingly the new design right would not give protection against industrial espionage of the type which occurred in *Hensher*.

The scope of the definition of design document is nevertheless quite wide. In *Squirewood Ltd v H Morris & Co Ltd* 1993 GWD 20-1239 the petitioners designed office furniture and produced brochures in which their furniture was described and illustrated by photographs and drawings. Lord Clyde granted interim interdict on the basis that the brochure was a design document. It could however have been made clearer in the opinion that design right did not subsist in the brochure as such or as a design document. Design right subsists in the design, of which the design document is a record only. It is not necessary to refer to any design document to determine whether infringement of a design right has taken place. As will be further discussed below (81-4), all that is needed for infringement is to establish the causal link of copying between the original design and subsequent commercial reproduction of articles exactly or substantially to that design (1988 Act, s 226).

The role of computer technology in design work is recognised in the definition of design document. Data stored in a computer is capable of constituting a design document. Section 215 of the Act also deals with the possibility of a computer generating a design without any significant human intervention. This is important in addressing the question of authorship of a design, which is the basic test for ownership of the design right. There are, however, three circumstances in which authorship does not lead to ownership:

(1) where a design is created in pursuance of a commission;

(2) where a design is created by an employee in the course of his employment; and

(3) where a design is computer-generated.

THE COPYRIGHT DESIGNS AND PATENTS ACT 1988  79

In the first two cases, design right belongs to the commissioner and the employer respectively. In the case of a computer-generated design, the design right belongs to the person "by whom the arrangements necessary for the creation of the design are undertaken" (1988 Act s 214(2)). Although this phrase is not wholly clear, it seems likely that the owner or possessor (say under a lease or conditional sale contract) of the computer will be owner of the design right. The situation should be distinguished from that where the human designer uses the computer as a tool towards the creation of the design, when it will not be computer-generated within the meaning of the Act and the other rules as to authorship, employment and commissions will apply.

## Length of Protection; Licences of Right

The term of design right is much shorter than that of either copyright or a registered design (1988 Act s 216). The clear policy is to encourage registration where that is possible. The basic formula is taken from that for the protection of topographies in the EC Directive on the subject. Design right lasts for fifteen years from the end of the year in which the design was recorded, or in which an article was made to the design, whichever of these is earlier in time. In addition, the right is restricted further if, within five years of the earlier of these two events, an article made to the design is made available for sale or hire anywhere in the world by or with the licence of the design right owner. The right will then expire ten years from the end of the calendar year in which the marketing took place. Accordingly, the fifteen-year period is a maximum which will be reduced by commercial exploitation of the design during the first five years of its existence. The use of two starting dates was explained thus by Mr Butcher (*Parliamentary Debates*, 14 June 1988, HC Standing Committee E, cols 586-7):

> The most important thing to note about the term of design right is that the people who are most concerned are not so much rights owners, but their competitors – the people who want to know when a particular product is no longer protected and is in the public domain. The problem with relying on the date of creation ... is that this is not likely to be known to anyone other than the designer. In other words that date is not a sufficiently public event. Marketing is however an essentially public event. In the vast majority of cases, competitors will first become aware of a design when products are marketed. They will therefore know with certainty the latest point at which the 10-year clock could have been started. ... Why then bring in the date of creation at all? ... The answer is that to avoid perpetual rights we must have some formula. If we relied only on marketing to start the countdown to the end of protection, designs which never reached the market would be protected by design right for ever and that cannot be right. We therefore have to put an upper limit on the design right term and the only way of doing that is to provide an upper limit counted from the creation of the design.

A further limit on the duration of design right is to be found in the provision of section 237 for licences of right to be obtainable during the last five years of the design right term. In effect, this means that the design right owner has an exclusive claim to the design only for five years after his initial exploitation of it.

The very short period of exclusive use caused some controversy in the parliamentary debates. It was criticised as inadequate to enable the right owner to recoup his development costs. The Government did not shift its position in response. There were other, external, pressures to be taken into account. In answer to a parliamentary question on 30 March 1988, the Secretary of State for Trade and Industry, Lord Young of Graffham, observed that "we have not specifically discussed the term of protection with the [European] Commission, our proposal was for 10 years and the Commission accepted that. Our view is that a longer term might well be struck down by the European Court" (30 March 1988, HL Question [495], cols 761-2). Later, at the Committee stage of the Bill's progress through the House of Commons, Mr Butcher stated that "it has been made clear to us that the European Commission would object if we increased the term or eliminated either of the exceptions" (14 June 1988, HC Standing Committee E, col 555). He amplified this at report stage (25 July 1988, HC, col 59):

> I was asked what evidence we had that the European Commission would object to our increasing the term of design protection or removing the exceptions. I accept that there is no published evidence in recent Commission documents or elsewhere, but we discussed the matter with the Commission in the run-up to the Bill. The Commission had already objected to the excessive protection for functional designs given by United Kingdom copyright and was equally concerned that any new design law should not move out of line with protection elsewhere in the Community. The Commission was satisfied with our proposals on this basis but made it clear that additional protection might well lead to a formal complaint under the European Community treaty. That does not amount to being dictated to by the European Commission, because 10 years' protection, with appropriate exceptions and licensing arrangements, represents what we regarded as necessary in any event.

The short term is clearly designed to foster the competitive environment in respect of unregistered designs generally; it contrasts sharply with the length of protection arising through registration and therefore encourages the use of that system. Indeed the Government argued that, given that inventive products could be protected for twenty years under a patent and that 'eye-appeal' designs could be protected for twenty-five years through registration, it had to be right that designs incapable of receiving such protection got a shorter term (*Parliamentary Debates*, 12 January 1988, HL, cols 1125-6; 14 June 1988, HC Standing Committee E, cols 587-90).

Criticism of the availability of licences of right within the last five years of the design right was met by the Government argument that such licences would of course be on terms either agreed by the parties or fixed by the Comptroller, and that the remuneration of the right owner would in either case take account of his investment and development costs. A statutory instrument made under the Act gives these as factors to be considered by the Comptroller in disputed cases, along with quality and safety (Design Right (Proceedings before Comptroller) Rules, SI 1989/1130).

## Infringement

Under section 226 the exclusive right is to reproduce the design for commercial purposes:

(1) by making articles to the design;
(2) by making design documents to enable such articles to be made.

Reproduction means copying the design, directly or indirectly, so as to produce articles exactly or substantially to that design. It is infringement to do these things without the licence of the design right owner. So design right, unlike the rights conferred by a patent or a registered design, is not a monopoly right; before there can be infringement of design right, there must be the causal link of copying between the two designs in issue. The requirement of copying before there is infringement is a limitation of design right by comparison with registered designs. Again the policy of encouraging registration is apparent.

The copyright test of copying will presumably apply and so, except for semiconductor topographies, 'reverse engineering' and 'redesign' will continue to be caught as infringements. Topographies have been taken to be a special case in this regard ever since the original legislation in the United States, since it is understood that most development in this area, as with much other computer-related technology, is based upon 'reverse engineering' in relation to existing products, and it is therefore regarded within the industry as a legitimate competitive technique rather than as an infringement of rights. Because design right escapes from the concept of artistic copyright and focuses simply on whether or not there is a design applied to an article, it should no longer be possible to run arguments over the mixture of artistic and literary elements in designs in order to limit the scope of what constitutes reproduction. There will still be difficult questions about exactly what reproduction is, however, since design continues to be defined primarily in terms of visual features.

In the *Klucznik* case the allegedly infringing pig fender had a rounded tube or roll bar on top but it differed from the first fender in having flaring sides which enabled it to be stacked with other fenders. The claim of infringement failed. Although there was substantial similarity in respect of the roll bars, the overall designs of the fenders were different. Aldous J argued that the test for infringement of unregistered design right was different from that for infringement of copyright. Whereas with copyright only copying of the work or of a substantial part of it had to be shown, with unregistered design right, it had to be shown that articles had been produced exactly or substantially to the design (s 226(2)). The judge went on ([1992] FSR at 428):

> Whether or not the alleged infringing article is made substantially to the plaintiff's design must be an objective test to be decided through the eyes of the person to whom the design is directed. Pig fenders are purchased by pig farmers and I have no doubt that they purchase them taking into account price and design. In the present case, the plaintiff's alleged infringing pig fenders do not have exactly the

same design as shown in the defendant's design document. Thus it is necessary to compare the plaintiff's pig fenders with the defendant's design drawing and, looking at the differences and similarities through the eyes of a person such as a pig farmer, decide whether the design of the plaintiff's pig fender is substantially the same as the design shown in the drawing.

Although it is probably right to say that the comparison between a design and an allegedly infringing article must be objective (see further Turner 1993), the "pig farmer" test appears to be an unjustified gloss on the statutory provisions. It is true that copying by producing articles to the design is not the same as copying a work in the copyright sense, namely, reproducing it or any substantial part thereof in any material form (ss 16(3) and 17(2)). But the real significance of the statutory words "making articles to that design" lies elsewhere in the 1988 Act, in section 51's exclusion of making articles to a design from the ambit of copyright in the design (see further below, 82-3). There is no reason to go on to test whether or not the article is made to the design by reference to the person to whom the design is directed. The question is simply whether a design has been copied – that is, whether there is a causal link between a design and a subsequent article. It was established that the designer of the second fender had seen the first one and had used it towards his own design (see [1992] FSR at 429). Aldous J had no doubt "that the idea of having a tube as the roll bar came from the defendant's pig fender and therefore copying did take place" ([1992] FSR at 429). That, it is submitted, should have been enough to establish a prima facie case of infringement of the design right in the first fender (see further Copinger & Skone James 1991: para 20-124). Indeed, to be more precise, given that only the roll bar was not of commonplace design in pig fenders, the infringement was of its design right rather than that of the fender as a whole (see above, 72, and Laddie Prescott and Vitoria 1995: para 43.5).

There is, however, a crucial contrast with copyright infringement: by virtue of section 51, making articles to a design is not an infringement of the copyright in the design. Right (i) above may otherwise be compared with the right conferred by registration of a design, which is to stop the unauthorised making of articles for sale or hire or use for business or trade purposes. The 'commercial purposes' which are necessary for there to be an infringement of unregistered design right must cover at least the same ground.

Right (ii) above appears to go further than the Registered Designs Act in giving exclusive rights to the reproduction of the design document. The 1949 Act does provide that the making of anything which enables an infringing article to be made is an infringement of the rights conferred by registration, but this is concerned only with three-dimensional items which could be used in the process of manufacturing infringing articles, such as moulds and print rollers for textiles, rather than with the making of copy design documents (although the language of the 1949 Act here does appear capable of covering this situation). It is also clear that there is an overlap of design right with copyright here, since section 51 does not take making a copy of a design document outwith the scope of copyright infringement. The importance of this is that a design document may have copyright (e.g. if it is a drawing), or it may not (e.g. if it is a model which is not a sculpture or work of artistic craftsmanship). Thus copying the design document as such may or may not be an infringement of copyright.

This, coupled with the operation of section 51, giving full copyright to designs for artistic works, and section 52, giving twenty-five years of copyright to articles manufactured from artistic works, underlies section 236 of the Act. Where copyright subsists in a work which consists of or includes a design in which design right subsists, it is not an infringement of design right in the design to do anything which is an infringement of the copyright in that work. In other words, the infringement action should be brought under the copyright provisions of the Act, and it is not to be assumed that because there has been infringement of copyright *ipso facto* there has been an infringement of the design right. It should also be borne in mind that infringement of copyright is in some respects much wider in scope than infringement of design right. Thus, for example, it will always be an infringement of copyright to make a copy design document so long as it has copyright in the first place, but the same act will only infringe design right if it is done for the purpose of manufacturing articles commercially. On the other hand, section 51 ensures that copyright is not infringed by the making of articles to the design. However where the design document does not have copyright, copying it will infringe design right if the purpose of the copyist is to manufacture articles commercially to the design.

A useful illustration of the interaction of copyright and design right can be found in the facts of *Squirewood Ltd v H Morris & Co Ltd*, in which Lord Clyde upheld the design right in office furniture and granted interim interdict. Because the decision was one on interim interdict, Lord Clyde's opinion contains little substantive discussion of unregistered design right. But the averments disclosed an interesting situation in which the interaction between unregistered design right and copyright was an important issue.

Squirewood designed office furniture. Their designs were contained in drawings and brochures which, as noted above (78), were held to be design documents. Squirewood claimed that Morris were manufacturing and selling office furniture which was in substance a reproduction of the designs. The grant of an interim interdict on the basis of both copyright and design right appears to have been correct so far as concerns the latter but only doubtfully so in respect of the former. Lord Clyde rested his decision on rights in the brochure. But design right does not subsist in a brochure as such or as a design document. Design right subsists in the design, of which the design document is a record only. It is not necessary to refer to any design document to determine whether infringement of a design right has taken place. As already indicated, all that is needed is to establish the causal link of copying between the original design and subsequent commercial reproduction of articles exactly or substantially to that design (1988 Act, s 226).

The brochure also had copyright. Lord Clyde held it to be a literary work also containing artistic works in the form of the photographs and drawings (transcript, p 2). But the copyright was almost certainly not infringed by the manufacture and sale of office furniture. While the copyright in a two-dimensional artistic work may be infringed by making a copy in three dimensions

(1988 Act, s 17(3)), this is limited by s 51 of the 1988 Act as part of the legislative policy of excluding industrial designs from copyright. To make furniture to someone else's design without permission is not an infringement of copyright unless the furniture is an artistic work in its own right. Furniture can constitute artistic work: as a sculpture or a work of artistic craftsmanship, for example. But this is not very likely when the work is intended for mass industrial production, as was probably the situation in the *Squirewood* case. The copyright in the brochure only prevented other parties from making two-dimensional reproductions of its contents.

## Remedies

When design right is infringed, the owner has all the usual remedies – damages, interdicts (injunctions in England) and accounting of profits. As with copyright, the court may award 'additional damages' in cases of 'flagrant' infringement, taking account of "any benefit accruing to the defender by reason of the infringement" (1988 Act s 229). This is a remedy which does not exist in the case of registered designs, although it was available for infringement of copyright under the 1956 Act. There, however, its scope was limited to cases where no other effective relief was available. Flagrancy was defined by Brightman J as "scandalous conduct, deceit and such like; it includes deliberate and calculated copyright infringements" (*Ravenscroft v Herbert* [1980] RPC at 208). The remedy was mainly applied where the act of infringement constituted an intrusion upon individual privacy. Thus in *Williams v Settle* [1960] 1 WLR 1072 and *The Lady Anne Tennant v Associated Newspapers Group* [1979] FSR 298, additional damages were recovered where the infringement of copyright consisted in the publication by newspapers of family or otherwise private photographs. Nonetheless additional damages were granted in at least one reported case of design copyright (*Nicholls Advanced Vehicle Systems v Rees* [1979] RPC 127 and [1988] RPC 73), while, when the *Interlego* case was in the Hong Kong Court of Appeal ([1987] FSR 409), the possibility of additional damages was discussed and rejected. The claim was made because there had been deliberate copying by the defendants Tyco. The court was satisfied, however, that this was not flagrant infringement since Tyco, acting on legal advice, had communicated their intentions to the Lego company and had not met with immediate objection. The 1988 Act removes the restriction that no other effective relief is available and it will be interesting to see how the remedy is now developed.

One limitation which will certainly continue to affect additional damages in design cases is the elimination of damages as a remedy where infringement is 'innocent', which arises under section 233, and is paralleled in both copyright and registered designs provisions. A defender will be innocent where he did not know, and had no reason to believe, that design right subsisted in the design to which the action relates. Such innocence will not affect the possibility of other remedies being obtained against him. However as a consequence of section 239 all the remedies available to a pursuer become very restricted in scope where the defender undertakes to take a licence: in particular, the right to an inderdict or injunction is lost. Again, there are parallel provisions for copyright and registered designs. It is thus extremely difficult to use infringement proceedings to prevent the design being used by competitors.

Another feature of copyright remedies which was much criticised in relation to industrial designs – by the Whitford Report, for example – was the absence of any protection for potential defenders akin to the claim arising under section 26 of the Registered Designs Act in respect of groundless threats of infringement actions. There is now a similar provision in respect of design right under section 253 of the 1988 Act.

## Reciprocity

A last important feature of design right concerns reciprocity. The model is once again the topography right invented in the United States in 1984 (above, 10). Under section 256, foreign nationals will only enjoy design right in the United Kingdom where their own legal system provides equivalent protection for British nationals. This meets the complaint, expressed with greatest force in the Trade and Industry Committee Report on the motor component industry, that foreign companies obtained a competitive edge by virtue of their access to British copyright law, enabling them to inhibit the activities of their British rivals in the United Kingdom, while being unable to claim such rights elsewhere in the world. The effect of the provision is that British companies are able to copy foreign products entering the United Kingdom without fear of litigation ensuing, unless they come from countries with equivalents to design right. Whatever the effect that this has had on employment in the manufacturing industries, it should ensure more work for comparative lawyers. The European Union countries have been recognised as granting an equivalent protection to the designs of British nationals, but not the United States and Japan (1988 Act s 217(3)(c); Design Right (Reciprocal Protection) Order, SI 1989/1294).

## Conclusions

Much of the policy which appears to have guided the formulation of the specific rules of design right apart from topography right is drawn from the law relating to patents and registered designs rather than copyright. This is evident in a number of features already discussed: for example, design right depends rather more on novelty than on originality in the copyright sense of independent production. Again, although design right is not a monopoly right, being dependent on copying, infringement is primarily the manufacture and marketing of industrial products, as with patents and registered designs, and copyright is for the most part carefully excluded from this area. The provisions on licences of right also have no parallel in copyright, although such licences are an important feature of patent law, and there are provisions for compulsory licences to be granted in certain circumstances in both patent and registered design law. A final aspect of the new right, also obviously taken from patent and registered design law and without parallel in copyright, is the right of the Crown to make use of an unregistered design without requiring authorisation from the owner, unrestricted despite the reservations on this matter of the Whitford Report.

It is apparent from this discussion that design right is hedged around with considerable limitations, often of uncertain scope, in respect of its existence, duration and remedies for infringement. The Government was clearly reluctant to accept any modification which had the effect of extending the right and, as a result, it is open to question how far, if at all, commercial enterprises will seek to make use of it, since it has the air of a fair-weather friend. Such reluctance will be reinforced by the conferral upon the Secretary of State of power to order the cancellation or modification of licences, or to order the provision of licences of right, in respect of designs, where the Monopolies Commission reports that the conduct of the design right owner in these matters may be expected to operate, or has operated, against the public interest (1988 Act s 238). Once again, similar provisions have been introduced for registered designs (1949 Act s 11A). This meets the problem encountered following the Monopolies Commission report on Ford, that there were no effective remedies available under the existing legislation to deal with the situation there disclosed. While this recognises that the essence of the whole problem in this area is the regulation of competition in the marketplace, where competitors are armed with such powers the exercise of rights becomes a matter fraught with a very high level of uncertainty.

# Chapter Five

## The European Future

In the first edition of this book it was forecast that the controversy over protection of designs as intellectual property would not cease as a result of the enactment of the Copyright Designs and Patents Act 1988 (MacQueen 1989: 81). Quite aside from the way the new rules would bed down as the courts interpreted them, the European Commission was known to be examining design rights. The 1988 Green Paper on copyright (para 1.6.3) made clear that the Commission regarded the subject as an issue with a bearing on the creation of a common market. On the wider international front, the draft TRIPS Agreement of the Uruguay Round of the GATT included provisions on industrial designs which offered for the first time the prospect of some international harmonisation on substantive issues of design protection.

The expectations of 1989 have been amply fulfilled. As the previous chapter has demonstrated, the 1988 Act has produced some difficult case law, although there has not been as much litigation as might have been expected and the most controversial case has been about words in the Registered Designs Act 1949 which were not amended in any way in 1988. In the absence of an international consensus on design protection, national laws have continued to go on their diverse ways (Berman and Lambrecht 1992; Firth 1993). The successful conclusion of the Uruguay Round in 1993-94 has meant, however, that there is now some prospect of harmonisation in this area. The World Intellectual Property Organisation (WIPO) is working on the development of the Hague Agreement concerning the international deposit of industrial designs, aiming "to facilitate, through one centralized procedure, obtaining protection in all the States party to the treaty" (WIPO 1995: para 5). The greatest controversy has attended the actions of the European Commission, which in 1991 produced a Green Paper on *The Legal Protection of Industrial Design*, followed at the end of 1993 by a draft Regulation (COM(93) 342 final), proposing Community registered and unregistered design rights, and a draft Directive (COM(93) 344 final), seeking the harmonisation of the national registered design systems on lines similar to the Community registered design.

In this chapter there will be first an appraisal of the developments in the United Kingdom under the 1988 Act, followed by a critique, in the light of that experience, of the draft Regulation and Directive produced by the Commission. Some comments will also be offered on the implications of the TRIPS Agreement for the United Kingdom and European laws. Finally there will be a return to the general issues about copyright and intellectual property raised in the opening chapter, and some concluding observations will be made.

## The Public Interest in Designs

Is the protection of designs by an intellectual property right in the public interest? A useful starting point in answering this question is a powerful analysis of the goals of intellectual property in general proffered by J H Reichman (1994), which argues that design right is a hybrid form of intellectual property between patent and copyright laws. Many such hybrids have sprung up in order to create a 'lead time' for producers in respect of goods which are ineligible for either patent or copyright protection and which, because they are also incapable for various reasons of being protected by trade secrets rules, are unlikely to have lead time in unfettered market conditions. Designs "bear their secrets on their face", being applied to products which are readily available in the market place, and if unprotected may be unable to earn the return justifying the cost of their creation from a commercial point of view. A dynamic market economy in which innovation and technological advances occur assumes a degree of natural lead time, and intellectual property of all kinds can cover for market failure in this respect. Reichman however deplores the proliferation of hybrid forms of intellectual property, and the lack of unifying principle which has informed their separate development, and reminds us that the classical bipolar paradigm of patents and copyright also has its 'negative' aspect, inasmuch as it assumes that many producers are not protected from imitation and consequent price competition; that disclosed material can be used by competitors; and that unfair competition law only represses imitation if customers are confused or deceived. If hybrid regimes are allowed to multiply, the "rising protectionist tide ... could so burden the innovative enterprise ... as to slow the pace and skew the direction of innovation more than would otherwise occur if free-riders were left alone" (Reichman 1994: 2533).

This argument clearly entails rejection of many of the values which underlay the rise of design copyright and its successor, unregistered design right, in particular the old notion that "what is worth copying is worth protecting". That is precisely the assumption which Reichman challenges. The fact that parties seek protection for their products because there are copyists or free riders about does not necessarily mean that they should get it. Reichman's (1994: 2533 ff) solution is "a general purpose innovation law capable of regulating publicly distributed embodiments of technical know-how that neither patent, copyright, nor classical trade secret law adequately protect ... [which] must provide qualified innovators with a modicum of artificial lead time, even after public distribution takes place, and not require secrecy, simply because competition presupposes lead time". It will unfortunately be some time before any such regime can be installed on a national or international basis, and we must meanwhile do the best we can with what we have, including the established hybrids and in particular design law. But in considering how the subject matter might be dealt with when there is, as at present in Europe, an opportunity to settle its framework, it would be useful to keep Reichman's comments in mind in thinking about what we are trying to achieve.

Where then does the public interest lie in relation to industrial designs? In various ways design has an important economic and social role. The need for good product design is a long-established and constant public theme. It is recognised as crucial to the market performance of goods. Well-designed products will do better, particularly overseas. In Scotland, Sir John Clerk of Penicuik recognised this as early as 1728, and in 1760 the Board of Trustees for Manufactures established an Academy in Edinburgh in which the teaching of pattern design for the linen industry was a principal object (Macmillan 1986: 44; Macmillan 1990: 93, 118). The message was again proclaimed at the Great Exhibition in 1851, and it has been carried forward since by organisations such as the Design and Industries Association, the Council of Industrial Design and, in our own time, the Design Council. Moreover, as people such as William Morris and Patrick Geddes were the first to show, design is significant for the general environment in which we live: to be surrounded by products which are well-designed, considered from both an aesthetic and functional point of view, seems both desirable and profitable. Today this message seems better understood than ever, both by individuals and by industry and commerce, although it is sometimes distorted by the tendency to use the word 'designer' as an adjective, "as a prefix for sales purposes, as 'added value', as a symbol of status and supposed quality" (Huygen 1989: 12). This may be an unintended consequence of a move away from the former orthodoxy that the aim of design was 'fitness for purpose' and that the form of an object should be dictated by its function. But there has been an ever-increasing awareness of the significance of good design, perhaps reflected in the rapid development of design consultancies. Some sense of the scale of the activity of these consultancies and, consequently, of the economic significance of the design industry (as distinct from design in industry) can be gained by glancing at a table of the top design fee earners in 1993 published in an issue of *Design Week* in March 1994. 81 consultancies had fee income of over £1 million and ten earned over £10 million. Probably relatively little of this was earned in product design, however, as the activities of such consultancies range through architecture, interiors, graphics, packaging and corporate image making (Huygen 1989: 67-9, 79-90). Nonetheless, the fact that design has become such a significant sector of the British economy suggests that adequate protection of its products is a matter of public interest, as providing a framework from which a return can be earned, which in turn will provide the incentive for further improvement.

The basic considerations for the legal response to this can be framed as follows. First, do design rights provide an incentive to useful activity which might not otherwise occur? – or, in Reichman's terms, with 'lead time' not available on the market? This has two aspects: one being the content of the right so that it is worth having, the other being the encouragement it gives to others to develop and improve upon the original idea, rather than simply permitting stagnation. Second, is there adequate access to the right? This also has two aspects: the success of the creator in terms of obtaining the right and the access of the public to his material within a reasonable time. Third, are there adequate controls upon the exercise of the right, so that if necessary the public interest can be asserted against individual self-interest?

## The Operation of the 1988 Act Appraised

Experience up to the *British Leyland* case showed that traditional copyright and registered design law were not appropriate forms of protection for all kinds of design. The law was complex. Much of it was inadequate in the protection which it offered, while other parts provided excessive protection and had some potential for affecting other aspects of the public interest adversely. The report of the Monopolies Commission (1985a) on the use of copyright by the Ford Motor Company to prevent competition in the car spare parts market made this clear, the consumer interest in being able to obtain lower prices for parts being the important counterbalance to the interest of the design originator to have exclusive rights. Such rights were not necessary to guarantee him his return on his skill and labour (or research and development). The cost of developing the design was recouped through sales of the whole vehicle rather than in the 'after-market' developing some years after it was first put on the market. In Reichman's terms, 'lead time' seemed to exist in the marketplace without the assistance of the law. Change in the law, the Commission concluded, was clearly necessary.

With the alternatives of registration or unregistered design right, the Government opted for a bifurcated approach to design protection. Of the two, registration is stronger but access is more difficult. The formal procedures, and the consequent requirement of professional assistance, are without doubt disincentives for many potential rights-holders. This was pointed out in the 1960s, and gave rise to the Design Copyright Act 1968; it was restated by the Whitford Report in 1977. In 1983 the Government's Green Paper stressed readier access to intellectual property rights as a key policy for future development. This bore rather limited fruit in September 1986 with the publication of proposals to break down restrictions in the practice of patent agents, the main professional advisers not only for patents but also for trade marks and registered designs (Office of Fair Trading 1986; Department of Trade and Industry 1986: 29-30). Some of these proposals were translated into provisions in the 1988 Act (ss 274-286). But these are not likely to provide an adequate solution to the problem. The annual figures on applications, registrations and renewals of registrations since the 1988 Act came into force suggest that extending the maximum period of protection to twenty-five years has not had a major impact in making registered design protection more attractive to industry (above, 67-8). There had already been a revival since the low point of the system in the 1970s, but while the numbers show that the registration system is useful and used to an appreciable extent, it is still well short of the levels of use achieved in the 1950s. The figures for renewals show that very few registrations carry on into a third five-year period (above, 68), and presumably even fewer will make use of the fourth and fifth periods which are now available. It is certainly very doubtful whether these additional periods have played any part in attracting designers to register rather than to rely upon unregistered design right. If again we translate this into Reichman's terms, the lead time which product designers seem to feel they need is rarely more than ten years and very often no more than five.

Further, the law retains the division of designs into those with eye appeal and those which are dictated by function, with the latter being unregistrable. While it is true that there is no need to protect absolutely every design created, this exclusionary criterion reflects an evaluation of what makes a worthwhile design which is essentially Victorian or earlier in origin, and it is surprising that it retains such a significant place not only in British law but in the legal systems of many other countries. While the 'eye appeal' concept is not an artistic one, it may be worth adding that the division between the aesthetic and the functional objects of everyday life is not one which has carried much conviction since Marcel Duchamp (1887-1968) presented urinals and bicycle wheels mounted on stools as works of art before the First World War. It is a theme made familiar today by, most famously, the soup cans of Andy Warhol (1930-1988), and by the presentation of piles of tyres and layouts of bricks as sculptures in galleries of modern art. More pragmatically, the distinction between eye appeal and functionality is difficult to apply and adds to the barriers fencing off registered design rights.

Further problems with the registration system have now emerged as a result of the decision about the limited meaning of the word 'article' in the *Ford* case in 1994. The decision of the House of Lords ([1995] 1 WLR 18) was clearly influenced by the hostility towards intellectual property rights in spare parts which has been evident since the report of the Monopolies Commission on Ford's spare parts policy. It is curious, however, that the House of Lords achieved its goal in the way it did rather than by making use of those parts of the legislation intended to limit or capable of restricting the registration of spare part designs – 'must match', functionality, and so on. Further, if the decision is correct, then a very large number of designs applied to the parts of motor vehicles were wrongly registered and protected under the 1949 Act before 1988. It seems certain that, if the decision is allowed to stand, there will be a decline in the use of the registration system.

Unregistered design right is more accessible than registered design right, and, as the *Klucznik* and *Squirewood* cases show, can clearly play a useful role in the protection of designs from misappropriation by competitors. But it can scarcely otherwise be described as a more attractive form of protection than registration, since it is limited in a variety of ways – term, scope, and subjection to compulsory licences and supervision by the Monopolies Commission. It is true that the latter two limitations also apply to registered designs, but licences can be obtained in respect of unregistered designs at any time during the last five years of the design right, whereas with a registered design there must be a failure to apply the design to a reasonable extent. In both cases the limitations are a recognition of the public interest as a balance to the claims of right-holders. One of the arguments of this book has been that intellectual property is not like any other forms of property and that the exclusive rights which it gives cannot be treated without regard to the public interest, which justifies them in the first place. The development of the role of the Monopolies Commission in the 1988 Act is its principal recognition of this point, while the licences of right for unregistered design right ensure both that there is competition and, since licences must be paid for, a contribution to the research and development costs of the initial producer. If there is still a consumer demand, then, the rightholder continues to benefit from it despite the existence of direct competition.

The length of the term and the spare parts exceptions to design right are also attempts to accommodate the public interest. The question of the length of term for any intellectual property right is always a difficult one and the choice to some extent arbitrary. The choice of ten years from first marketing, limited by a ceiling of fifteen years from creation, seems to be about right, especially given the length of protection provided by other forms of intellectual property, in particular patents. This conclusion is reinforced, it is suggested, by the number of renewals of registered designs, which show that very few indeed are protected for longer than ten years, while a majority are not protected for more than five. This indicates that the lead time provided by design protection certainly should not exceed ten years and could probably be much shorter without doing serious damage to producer returns in all but a few cases.

On a number of grounds, however, the spare parts exception is harder to justify than either the length of the term or the availability of licences of right. First, there is the sheer difficulty of the exception, which produces considerable uncertainty. Accordingly, it will always be difficult for the manufacturer of any machine made up of components to know whether or not he has intellectual property rights capable of protection. If value is placed on access to these rights, then the spare parts exception detracts from that. The arbitrary nature of the exceptions is evident in the decisions of the lower courts in the *Ford* case, under which some parts survived to receive protection while others fell by the wayside. Second, the exception was plainly formed with the situation in the motor industry at the forefront of the draftsman's mind. The 'must fit' and 'must match' exceptions seem to stem almost straight from *British Leyland v Armstrong Patents* and the Monopolies Commission's Ford report. Although the case law showed that copyright in spare parts was of importance in other industrial sectors, notably engineering, there was little apparent investigation of the market situation outside the motor industry. Clearly, as the responses to the various reports and papers on the subject showed, there were deep divisions of opinion across British industry on the subject. One might have thought that the best way to respond to this was to avoid legislation plumping so firmly for one side of the argument. What is worse is that the courts will be deciding these issues, which are essentially about economic activity and commercial competition, in accordance with laws rather than in accordance with what the situation in the industry concerned or, where relevant, with what the consumer interest may require.

The recognition that intellectual property rights are about market power is essential to a true understanding of the law. But one should not fall into the error of assuming that, because the right-holder has a potential monopoly, it follows that he can control the market. This is nicely illustrated by a series of cases since the mid-1970s, about the publishing and recording contracts entered into by recording companies with popular singer/songwriters and groups such as Fleetwood Mac, Elton John, Gilbert O'Sullivan and, most recently, George Michael. The cases have essentially been about whether the contracts were fair or not, and in most the famous name was able to have them struck down on one ground or another. Even though he or they were the owners of the copyright in their material, their bargaining power in relation to their licensees or assignees, the recording companies, was very limited in negotiating for their publication. With the notable exception of the George Michael case, the courts have generally been of the view that the musicians needed greater protection from the imposition of unfair terms than they had received. Critics have argued for and against the proposition that the contracts in question *were* fair, and that the profits which the recording companies made from the success of certain individuals were ploughed back into giving many others the opportunity to follow in their footsteps, more than would otherwise be possible (Trebilcock 1976; Atiyah 1986: 155-6; Monopolies Commission 1994: 97-100). Whatever the truth of the matter, the cases demonstrate how complex the question of market power and its effects may be, and how unlikely it is that, by looking at one case in isolation or out of context, as tends to be the way in litigation, a decision will be well informed on the general public interest.

There is here a genuine difficulty, which is well evidenced by the *British Leyland* case. There are frequent references in the speeches to the way in which BL was exercising a monopoly which prevented the consumer from obtaining spare parts for his car, or which made him pay more for his parts than would otherwise be the case. Yet the evidence in the case, particularly as it emerges in the judgments of the lower courts, was that BL was supplying the market and that its licensees were able to compete in a way to which BL had to respond. Some of the language which the Law Lords used suggested that they had more in mind the policy of Ford as disclosed in the Monopolies Commission report the previous year. The truth of the matter is that an adversarial court procedure is not well suited for the resolution of the types of issue which arose in the *British Leyland* case. The inquisitorial and investigative techniques of the Commission seem more appropriate. Lord Bridge hinted at this kind of difficulty in his speech in *British Leyland v Armstrong Patents* [1986] AC at 626 when he referred to "the problem as to where, if at all, and if so by what criteria, the law can draw a line to discriminate between acceptable and unacceptable claims to enforce copyright which restrict the market in spare parts". In his speech in *CBS Songs Ltd v Amstrad Consumer Electronics plc* [1988] AC 1013 Lord Templeman seemed also to realise the limitations of the law's rule-based approach and the adversarial nature of court proceedings in meeting such problems adequately. The case, it will be remembered, concerned home taping of sound recordings as an infringement of copyright. The court had been presented with estimates that, if the use of fifty million blank tapes could be prevented, some thirty million more records would be sold and the profits of the recording industry would greatly increase. That might not be so, but Lord

Templeman accepted that there must be millions of breaches of copyright by home taping every year. "Whatever the reasons for home copying, the beat of Sergeant Pepper and the soaring sounds of the Miserere from unlawful copies are more powerful than law-abiding instincts or twinges of conscience" ([1988] AC at 1060). Lord Templeman thought it lamentable for society that the law should be so treated . Nonetheless ([1988] AC at 1060):

> In these proceedings the court is being asked to forbid the sale of all or some selected types of tape recorder or to ensure that advertisements for tape recorders shall be censored by the court on behalf of copyright owners. The court has no power to make such orders and judges are not qualified to decide whether a restraint should be placed on the manufacture of electronic equipment or on the contents of advertising.

Lord Templeman went on to call upon Parliament to change the law in appropriate ways. But he might also have called for an investigation of the market situation upon which appropriate law reform might be based. It was certainly clear that litigation was not the best way to take account of all the interests concerned.

When we consider that the spare parts exception is designed to ensure competition in the public interest, and recall that design right is already subject to the control of licences of right and Monopolies Commission supervision to protect the public interest in competition, it seems a fair conclusion that there are too many tools here, all to do the same job. For the reasons just discussed, the spare parts exception seems the least satisfactory and most inflexible of these. Of all the ways to prevent unfair monopolistic behaviour, litigation is the least attractive precisely because it is a kind of contest between the parties and puts regulation of the market in the hands of the participants – and, we might add, the participants with the deepest pockets. For this reason, to leave the area to a general law of unfair competition, as suggested by Christine Fellner (1985: 199-203), seems uninviting. It makes litigation, or the threat of litigation, a competitive technique and would subject trading practice to monitoring only at the hands of business rivals. Self-regulation is fashionable and, in some respects, a valuable technique but not, it is submitted, in this form, which tends to be unable either to assess the public interest adequately or to take it into account at all.[1]

In general, then, it may be suggested that the British structure of design right, while recognising the balance to be struck between the creators and the consumers of intellectual property, is too complex. There are now three possible methods of protection in copyright, registration and unregistered design right. Furthermore, the last of these three is almost schizophrenic as rights given with one hand are taken away or minimised by the other. There must nonetheless be some sympathy for the Government which promulgated the legislation. In approaching this question, it received a mass of conflicting advice and was subjected to a variety of pressures, to some of which it had no option but to submit. For example, the survival of the registration system, despite the Government's commitment to easier access to intellectual property rights, owed much to a sense that this was the favoured form of design protection within the European Commission – a sense which, as will be seen below, has turned out to be well justified. At the same time a demand for design protection is plainly present in varying forms within the industrial and commercial communities. The evidence given to the Trade and Industry Committee investigating the British motor industry (1986-87: 177, 204, 215, 269, 280) suggested that, while design rights were not regarded by any means as the sole incentive for product and design innovation – response to market pressures and competition was at least as important – nevertheless they were valued and helpful.

## EC Proposals for Design Protection

The need for a European Community initiative on design protection essentially grew out of the very variable rules and systems on the subject to be found in the Member States. The jurisprudence of the European Court in the 1980s made it increasingly clear that holders of national intellectual property rights in one Member State could use them to stop the movement of goods from another Member State in which there was no equivalent right or where any right there was had expired. Thus the sheer variety of national design laws posed a direct barrier to the free movement of goods and the creation of a single market. Although in 1988 the Commission indicated that design law was low in its list of priorities (above, 87), only three years later there was a Green Paper on the subject; and this was followed at the end of 1993 by the publication of drafts for a Regulation establishing a Community design right (both registered and unregistered) and a Directive harmonising the national laws of the Member States. If all goes smoothly, the Regulation and Directive may take effect in 1996 or 1997.

What will be the effect on the law of the United Kingdom? The draft Directive is to be without prejudice to any system of unregistered design rights such as exists in the United Kingdom. Design is defined as the appearance of the whole or a part of a product resulting from the specific features of the lines, contours, colours, shape and/or materials of the product itself and/or its ornamentation. A product is any industrial or handicraft item, including parts intended to be assembled into a complex item, sets or compositions of items, packaging, get-ups, graphic symbols and typographic typefaces; but computer programs are excluded. Thus spare and replacement parts for other products can in principle be the subject of a registered design – a reversal of the position now reached in the United Kingdom as a result of the *Ford* case in 1994. There will be no need for the design to be applied to a set number of products, so the United Kingdom rule requiring application to at least fifty articles will disappear. Designs may also be protected by copyright, although the extent of such protection can be determined by each Member State. The tests of registrability will be novelty, judged on a world-wide basis, and possession of an "individual character", meaning that it produces a significantly different overall impression upon the informed user by comparison with any previous design. In assessing novelty and individual character, no account will be taken of publication or use by or through the designer during the twelve months preceding the filing of an application (a 'period of grace'). This will entail an adjustment to the Registered Designs Act 1949 and remove the difficulty about market-testing a design before undertaking the expense of registration (above, 31). Functional designs, defined as ones where the realisation of a technical function leaves no freedom as regards arbitrary features of appearance, are to be excluded from protection. This definition of functionality appears to be similar to that which prevailed in the United Kingdom before the *Amp* case in 1972 (above, 30); if this is so, then the number of registrable designs should increase significantly. There is no requirement of "eye appeal" or an "aesthetic" element. The design of a product which constitutes part of a complex item shall only be considered to be new and of an individual character so far as the design applied to the part as such fulfils those criteria. There is however a "must fit" exclusion from protection: "a design right shall not subsist in a design to the extent that it must necessarily be reproduced in its exact form and dimensions in order to permit the product in which the design is incorporated or to which it is applied to be mechanically assembled or connected with another product" (art 7(2)). Additionally there is a limited sort of "must match" protection: three years after the first marketing of the product to which a protected design has been applied, third parties may reproduce the design to effect repairs of complex products of which the protected product is part and upon the appearance of which the protected design is dependent. Protection is to last for five years in the first instance but will be renewable for five-year periods up to a maximum overall of twenty-five years; the same as the present position in the United Kingdom. The right conferred is a "monopoly" one of exclusive use in the marketplace rather than merely a right to prevent imitations, again the same as under the 1949 Act; but whereas the United Kingdom right is only in respect of articles for which the design has been registered, the Directive states that infringement is use of a product in which the design has been incorporated or to which it has been applied. Infringing use under the Directive includes exporting and stocking such products, neither of which is caught under the Registered Designs Act 1949.

The draft Regulation is a necessarily more complex affair involving the creation of both a registered and an unregistered design right at Community level. It therefore requires the establishment of a Community Designs Registry, which will be located with the Community Trade Mark Office at Alicante in Spain. The definition of, criteria for, and scope of protection under the Community registered design are much as those laid down for national systems under the draft Directive. The Community unregistered design has the same definitions and criteria for protection apart from registration, but the scope of protection is less. The spare parts and repair provisions apply to Community designs generally. There will be a one-year period of grace for use and disclosure of a Community design prior to filing for registration. Here there is a complex interaction with the Community unregistered design right, which in a model seemingly derived from the US Semiconductor Chip Protection Act 1984 (see above, 59) will last for only three years from the time at which the designer first makes the design available to the public. It is intended to provide protection for designs with short commercial lives where registration will not be worthwhile – for example, in the fashion industry – and to give designers an opportunity to put a new product on the market to see whether registration of its design would be useful; it will also protect designs intended for registration but which have not yet completed the process. When an unregistered design is put on the market, however, the designer has to be aware that failure to file within twelve months thereafter runs the risk of losing novelty and individual character, and thus the ability to register. Since the Community unregistered design will also only apply to designs which are not functional, albeit in a much more restricted sense than the current understanding in this country, it will also offer less coverage than the unregistered design right of the United Kingdom, which extends to designs dictated by function so long as they are neither "must fit", "must match", or commonplace. The three-year period of protection for the Community unregistered design is significantly shorter than that provided for unregistered designs in the United Kingdom, which is the lesser of fifteen years from creation or ten years from first marketing. Three years also seems rather too short to be generally useful, although it may meet Reichman's criterion of providing lead time sufficient to generate a return justifying the investment in creation, at least in industries affected by rapidly changing fashion. But like the British unregistered design right, the protection will not give an absolute monopoly over commercial reproduction of the design, but will instead prevent only copying of the design.

From a procedural point of view, it should be easier to obtain a Community registration than a United Kingdom one. Whereas a United Kingdom application must specify the article or set of articles to which the design is to be applied, the Community application *may* contain indications of the products to which the design is to be applied, or a classification of the products to which the design is to be applied; further, several designs may be combined in a multiple application. Again, whereas in the United Kingdom there is a substantive examination of the design for novelty and registrability, under the Community system only if a design is obviously not registrable will registration be refused. This should considerably reduce the time and costs involved in obtaining a Community registration.

It is very evident from all this that the European Commission favours a registration system as the principal means of protecting industrial designs. In this it is following the recommendations of the Max Planck Institute of Munich, whose 1990 study of design law was the basis for the Green Paper in 1991. "The purpose of registration is to create legal certainty as to which designs are protected and which are not" (European Commission 1991: para 4.3.9). This has been a constant theme of arguments in favour of registration, and certainly in the context of a marketplace as huge as the European Union it can be readily seen how difficult it could otherwise be to be sure that no other business is using a design like the one that has just been created and gone to market. On the other hand, this is only an issue where design protection is conceived as a monopoly right giving the ability to control even independent production. If, as is the case with the existing British and the proposed EC unregistered design, the protection is only against imitation, then a designer need not be concerned about designs already on the market somewhere but of which he is unaware, because his work, not being a reproduction, will not infringe the existing right. The desire for certainty also tends to be most strong amongst large businesses, which can of course afford the time and cost of registration. If the aim is really as so often stated by the Commission, that is, to offer small and medium-sized enterprises convenient and low-cost access to intellectual property protection, unregistered protection along the lines of the British model seems, in the light of the cases to date, to go furthest in that direction, despite the efforts to make Community registration quick and inexpensive.

A further question hanging over a Community Registered Design is whether there is a sufficient demand for it. It was noted earlier (above, 31) that in the United Kingdom there are many times fewer applications for protection and registrations of design than with patents and trade marks. This pattern is found in other countries in Europe and elsewhere, along with the striking fact that everywhere, albeit to varying degrees, it appears to be essentially a protection for domestic rather than foreign industry. Phillips (1993: 436) comments that "while the absence of an attractive and powerful international filing system is arguably an important factor [*to explain a lack of foreign applications*], the question may be asked as to whether there is sufficient demand for international protection to justify any change in the present system". It certainly raises doubts as to whether businesses will take up Europe-wide design rights which can only be obtained by means of registration. Nor can one say that the possibility of twenty-five years of protection with a Community registration – longer than anywhere in the European Union apart from the United Kingdom and France (Firth 1993) – is likely to prove such an attraction to overwhelm other doubts, since the evidence from the United Kingdom itself is that a large majority of registrations are not preserved beyond their tenth year at latest. Another problem may be that the registration can be revoked within a particular territory by its national court, thereby losing its Union-wide character.

At least some critics favour an approach through a copyright or a copyright-style system of automatic protection (see Cornish 1991). Prominent amongst them is the Dutch scholar Herman Cohen Jehoram (1992; 1994b), who argues that design, as an expression of human imagination rather than the invention of a technical effect, is fundamentally a copyright matter. The low use of registration systems means that already in Europe most designs are either protected by copyright or unfair competition rules or not at all. Both the draft Regulation and Directive permit existing national forms of protection outside registration, in particular copyright, to continue in existence, although these are highly variable around the Union. Therefore it is argued that what is needed, whether or not there is a European registration system, is a harmonisation of the protection of unregistered designs on a copyright basis. Cohen Jehoram appears to favour a fairly undiluted model of copyright for this harmonisation, and suggests that the basis for it already exists in the general copyright harmonisation programme of the Union (above, 13). He is accordingly sharply critical of the Community unregistered design right proposal – 'a tepid hybrid' (Cohen Jehoram 1992: 76).

However, it has to be said that the United Kingdom experience suggests that copyright is an inappropriate form of protection, in some respects too strong, in others too weak. It is not just a matter of the excessively lengthy term of copyright, which is wholly unnecessary to provide the requisite lead time, but also the inability of copyright to deal with all the media in which design work is carried out. It is also the case that today industrial design is not just the simple visual idea which has dominated the development of the law until now and which underpins the possible applicability of copyright to the subject. Design of a product involves much more than simply the determination of its shape or configuration, or of how it is to be ornamented. It involves questions of choice of materials, of weight and 'feel', of processing and technology, and of the many other factors which go into the creation of a product which will perform its function and attract custom. Appearance may be completely irrelevant to technical designs such as those for electronic or mechanical circuits and chemical processes. The Explanatory Memorandum which the European Commission (1993: 10) issued with the draft Regulation states that its definition of design includes "any feature of appearance which can be perceived by the human senses as regards sight *and tactility* ... Weight and flexibility, for example, may in some cases be design features ... A material or a texture can likewise be the expression of a highly original idea and be a decisive element in perceiving the presence of a protectable design". The wording of the definition of 'design' in the draft Regulation and Directive does not appear to be apt to embrace these tactile features, starting as it does with the word 'appearance'; but no doubt that will be amended as necessary. The important point is that if design protection extends as far as this, then copyright cannot be the appropriate form of protection.

Predictably, the greatest controversy has been sparked by the treatment of the spare parts issue in the two drafts (Horton 1994: 53-6). The Max Planck Institute (1991) draft proposals contained no specific provision for "must fit" or repair exceptions to design right; the only relevant exclusion would be the limited one for functionality. Competition issues were to be treated by competition rather than intellectual property law. The Commission Green Paper (1991) introduced the "must fit" exception, and the repair exception appeared for the first time in the draft Regulation and Directive. The "must fit" proposal was criticised by Cohen Jehoram (1992: 76): "Why is protection withheld from interfaces in such a sweeping manner? In fact, it is only necessary to avoid certain particular abuses, like those experienced with spare parts in the car industry. This could be dealt with by much narrower wording". A much more comprehensive attack has been launched against the draft proposals by Professor Friedrich-Karl Beier (1994) of the Max Planck Institute. In an opinion drawn up on behalf of the European Automobile Manufacturers Association, he argues that "must match" exceptions "contradict the essence of design protection", because "a shape dictated by aesthetic reasons is the very incarnation of a design worthy of protection" (Beier 1994: 851). The "must fit" provision is unnecessary given the exclusion of functional designs. It is inappropriate to introduce competition considerations into intellectual property rights, and the exceptions are inconsistent with the rulings of the Court of

Justice in the Renault and Volvo cases. The exceptions encourage competition by imitation rather than innovation, which is economically undesirable. The industries promoted will be those of imitators in the low-cost countries of south-east Asia rather than European ones.

In the light of the comments already made on the spare parts exceptions in British law, it is hardly necessary for me to indicate broad agreement with the thrust of Professor Beier's critique. Certainly the experience of the *Ford* case in Britain does not suggest that exceptions for spare parts will prove easy to interpret and apply, even in a case where they are obviously intended to bite. However, the comment about competition law may require qualification in the light particularly of the Court's decision in *Magill* (above, 21-3), while it cannot be said to be altogether surprising that spare parts exceptions have now emerged in design law proposals, given the long-established hostility of the Commission towards manufacturers' attempts to tie consumers in to the purchase of spare parts, both with or without the use of intellectual property to that end. But to attempt to write this hostility into exclusionary legal rules seems misguided, for reasons already given (above, 92). Finally, while it may be true that competition by innovation is better than competition by imitation, Reichman's observation that intellectual property has never guaranteed a comprehensive protection against imitation needs to be kept in mind.

The overall assessment of the European proposals on design rights must nonetheless be one of fundamental doubts: about the need for registration, the duration of both the registered and the unregistered right, and the spare parts exceptions. Within a registered design framework, some aspects of the Directive might well make British law more useful to industry: the re-establishment of parts as articles, a return to the former very narrow definition of functionality, the removal of the absurd eye appeal test, and the introduction of a grace period are particular examples. But it would seem better overall, if the need to protect design is accepted, to pursue an unregistered design right as the basis for future European protection. The British scheme could provide a model in some respects: ten years of protection after first marketing, subject to licences of right in the last five, should ensure enough lead time for most designs, as well as competition thereafter which nonetheless ensured that competitors made some contribution to the costs of developing the design in the first place. Indeed this might be the best solution to the problems of the car industry, since it is commonly said that the spare parts market really comes into existence roughly five years after the vehicle is first put on the market. Thus independent suppliers of spare parts would be able to enter the market and compete with the vehicle manufacturers, who would also be unable to dictate absolutely the terms and conditions under which competition took place.

## International Framework

There is nothing radically inconsistent with the suggested model in the international framework for the substantive law which now exists in the shape of the TRIPS Agreement. Articles 25 and 26 of the Agreement provide that States shall give protection for independently created designs which are new or original. It is not required, however, that this protection shall be through a registration system. States may provide that protection is not to extend to designs dictated essentially by technical or functional considerations; the language is permissive rather than mandatory. The right conferred is to prevent commercial use of "copies" of the design, perhaps hinting at a copyright rather than a "monopoly" form of protection against infringement. Limited exceptions to the right are allowed, so long as these do not conflict with normal exploitation of design rights and do not unreasonably prejudice the owner's legitimate interests, taking account of third parties' legitimate interests. This therefore leaves the question of spare parts open, in language akin to Article 9 of the Berne Copyright Convention. Professor Beier (1994: 868-9) has argued that the "must fit" and "repair" exceptions of the European proposals are incompatible with TRIPS, since they in effect abolish rights of exploitation, "and therefore interfere with the interests of the design right owner in an unreasonable manner". The same cannot be said of a licence of right, since it does not prevent the owner exploiting his rights himself, it ensures that he is paid for use of the design by others, and the legitimate interests of consumers – for example, to repair their goods – are protected. The duration of protection under TRIPS is to be at least ten years, which as already suggested seems to be a sensible maximum period for designs from a commercial point of view. Finally, special provision is made in Article 25 for textile designs: States are not to impair designers' opportunity to seek and obtain protection, in particular through cost, or requirements of examination or publication, and this may be done through industrial design or copyright law.

## Conclusions

In conclusion, the history of design protection in this century illustrates well the argument advanced in the first chapter of this book. Conflict can arise between copyright and competition considerations when the former is too widely applied; copyright can be used as a brake upon the otherwise legitimate operations of the market. Copyright and other intellectual property rights are justified in many ways, and they play a significant and generally unchallengeable role in several areas of the modern economy. But it is clear that occasionally the exercise of copyright is inappropriate: industrial design is an instance of this. The extension of copyright to this field in the United Kingdom resulted from the view that it is wrong to copy another man's work, which itself arises from the notion that the act of creation or origination gives rise to rights akin to those of property. This is again characteristic of the modern development of copyright in general. Many things which should have told against the intrusion of copyright into industrial designs were overlooked. On a formal basis, there was the general legislative policy in the area, which was plain enough despite the obscurities of the language in which it was couched. The policy of confining protection to certain types of products and designs might have been misguided, founded as it was on outmoded concepts of product design; but that was not an issue for the courts to deal with, as it was firmly lodged in the legislation. Less formally, it was not at all clear that copyright law provided an appropriate form of protection for design work, being a most curious mixture of excessive strength and extreme weakness. But, in the pragmatic tradition of the courts, copyright was made to work. Then there were the socio-economic questions about the effect of this development of copyright upon the markets of Britain. This aspect of copyright has tended to be played down in discussions of the subject, so perhaps courts and legislators can be forgiven for overlooking it in the early development of design copyright. So far as design goes, however, the legislature has learned the lesson and introduced controls upon the exercise of intellectual property for the sake of the public interest, even if we may disagree with some of the solutions given effect in the 1988 Act. There is also a more sceptical judicial approach to copyright claims, even though in some cases it has been taken too far. We are less likely than was once perhaps the case to see copyright being manipulated to ensure that imitators receive their just deserts.

The history of design copyright suggests that other applications of copyright might also be examined with a more critical eye than has always been evident hitherto. Computer programs provide a prime example where copyright has been made to fit, albeit with much discomfort and unease; compilations and databases may be another. Again, we may acknowledge that the term of copyright for literary and artistic works is reflective of cultural as well as economic considerations, but what is the justification for providing the producers of works such as sound recordings, films and broadcasts with nearly as lengthy copyrights? To what extent is copyright being applied here only to ensure that producers of such easily reproduced works have lead time enabling them to reap the reward they need to keep producing? If that is the case, is a period of protection lasting fifty years or more really needed to achieve this purpose? Competition is not the only element in the public interest which may be relevant. Another aspect might be education; copying for educational purposes is allowed only subject to severe restrictions under the 1988 Act, the result of successful lobbying by educational publishers. The effect is to create difficulties for teachers and students and perhaps to affect the quality of educational provision. Where does the balance of interests lie between authors and publishers, on the one hand, and users of their work on the other? The material assembled and reviewed in this book shows that, in the argument between originator and copyist or user, all is not necessarily clear-cut and that the public interest does not invariably favour the former. When problems arise as a result of the clashes of interest involved, solutions are far from obvious.

The problems of design right have been resolved, for the time being, by taking designs out of copyright and making special, *sui generis* provision. This recognises that the difficulties cannot be solved by the application of general principles of copyright law, which are intended to cover many other areas of activity and have features inappropriate or inadequate for the protection of designs. This hiving off of a particular subject from the body of copyright, reminiscent of the nineteenth-century approach whereby there was a variety of rules for different fields of work, may be an indicator of the breakdown of the general concept of copyright. On the other hand, it may be that the discussion of unregistered design rights in the United Kingdom and now in the European Union points the way forward to a model for a general system which provides appropriate protection of the innovation which cannot be protected by copyright, patent or trade secret laws. It provides lead time after distribution of no more than ten years; licences of right permit competition after an even shorter period of time but ensure that competitors contribute to the costs of production unless they innovate; and competition authority scrutiny should therefore only be necessary in very exceptional circumstances (Reichman 1994: 2534). We may be on the brink, not of proliferation of yet further hybrid forms of intellectual property, but of a great generalisation as significant in its way as the development of copyright itself.

# Notes

### Chapter One
1. Plant (1934: 182) notes evidence that at the time the 1842 Act was also understood to be for the benefit of the family of Sir Walter Scott.
2. However the Department of Trade and Industry (1989: para 2.30) did state that in any reform of restrictive trade practices legislation intellectual property rights should be excluded from its ambit on the grounds that "the restrictions in the grant of a right to use property do not restrict competition, since in the absence of the licence there would be none". Later proposals for reform stated that there was no need to deal with intellectual property since restrictive licensing was covered by its own legislation (Department of Trade and Industry 1992: para 2.16).
3. The Monopolies Commission is now (April 1995) investigating the Performing Right Society.

### Chapter Two
1. The Patent Office has also been an Executive Agency since 1990, producing an operating surplus of £6.6 million in 1993-94.
2. It is of interest to note that in August 1994 the Italian competition authority ruled that the exercise of intellectual property rights in spare parts for cars constituted a monopoly which was against the public interest ([1995] 3 EIPR D-71).

### Chapter Five
1. This is not to say that in other respects the introduction to the United Kingdom of a law against unfair competition, such as was unsuccessfully proposed during the passage of the Trade Marks Bill through Parliament in 1994, would necessarily be a bad thing.

# Bibliography

Adams, J., (1987). *Merchandising intellectual property*. London: Sweet & Maxwell.
Annand, R., and Norman, H., (1994). *Blackstone's guide to the Trade Marks Act 1994*. London: Blackstone Press.
Atiyah, P. S., (1986). *Essays on contract*. Oxford: Clarendon Press.
Beier, F-K., (1994). Protection for spare parts in the proposals for a European design law. *International Review of Industrial Property and Copyright*, **25**, 840-879.
Bennion, F. A. R., (1984). *Statutory interpretation*. London: Butterworths.
Bently, L., (1994). Copyright and the death of the author in literature and law. *Modern Law Review*, **57**(6) 973-986.
Berman, C., and Lambrecht, N. S., (1992). Designs in the United States and Japan. *European Intellectual Property Review* **2**, 37-48.
Boswell, J, (1774). *The decisions of the Court of Session upon the question of literary property in the cause John Hinton of London, bookseller, against Alexander Donaldson and John Wood, booksellers in Edinburgh and James Meurose, bookseller in Kilmarnock*. Edinburgh.
Breyer, S., (1970). The uneasy case for copyright: a study of copyright in books, photocopies, and computer programs. *Harvard Law Review*, **84**(2) 281-351.
Breyer, S., (1972). Copyright: a rejoinder. *University of California at Los Angeles Law Review*, **20**, 75-83.
Carlyle, T., (1839). Petition on the copyright bill. *The Examiner*, 7 April.
Cohen Jehoram, H., (1992). The EC Green Paper on the legal protection of industrial design: half way down the right track – a view from the Benelux. *European Intellectual Property Review*, **3**, 75-78.
Cohen Jehoram, H., (1994a). The EC copyright directives, economics and authors' rights. *International Review of Industrial Property and Copyright*, **25**, 821-839.
Cohen Jehoram, H., (1994b). Cumulation of protection in the EC design proposals. The Herchel Smith Lecture 1994. *European Intellectual Property Review*, **12**, 514-520.
Copinger and Skone James (1991). *Copinger and Skone James on copyright*. 13th ed. London: Sweet & Maxwell.
Cornish, W. R., (1981). *Intellectual property: patents, copyright, trade marks and allied rights*. 1st ed (2nd ed 1989). London: Sweet & Maxwell.
Cornish, W. R., (1991). Designs again. *European Intellectual Property Review*, **1**, 3-4.
Cornish, W. R., (1995). Authors in law. *Modern Law Review*, **58**(1), 1-16.
Davies, G., (1994). *Copyright and the public interest*. IIC Studies in Industrial Property and Copyright Law. Max Planck Institute for Foreign and International Patent, Copyright and Competition Law, Munich: VCH.
Department of Trade and Industry (1981). *Reform of the law relating to copyright, designs and performers' protection*. Cmnd 8302. London: HMSO.

Department of Trade and Industry (1986). *Intellectual property and innovation.* Cmnd 9712. London: HMSO
Department of Trade and Industry (1989). *Opening markets: new policy on restrictive trade practices.* Cm 727. London: HMSO.
Department of Trade and Industry (1990). *Reform of trade marks law.* Cm 1203. London: HMSO.
Department of Trade and Industry (1992). *Abuse of market power: a consultative document on possible legislative options.* Cm 2100. London: HMSO.
Dixon, A., and Self, L., (1994). Copyright protection for the information superhighway. *European Intellectual Property Review*, **11**, 465-472.
European Commission (1988). *Copyright and the challenge of technology.* COM (88) 172 final. Brussels.
European Commission (1991). *The legal protection of industrial designs.* Brussels.
European Commission (1993). *Explanatory memorandum on the proposal for a European parliament and council regulation on the community design.* COM(93) 342 final. Brussels.
Feather, J., (1988). *A history of British publishing.* London: Routledge.
Fellner, C., (1985). *The future of legal protection for industrial design.* Oxford: ESC Publishing Ltd.
Firth, A., (1993). Aspects of design protection in Europe. *European Intellectual Property Review*, **2**, 42-47.
Gregory Report (1952). *Report of the copyright committee.* Cmd 8862. London: HMSO.
Greig, J. Y. T., (1969). *The letters of David Hume* (2 vols, reprint of 1932 ed). Oxford: Clarendon Press.
Haines, S., (1994). Copyright takes the dominant position. *European Intellectual Property Review*, **9**, 401-403.
Hill, G. B., and Powell, L. F., (1934-64). *Boswell's life of Johnson together with Boswell's journal of a tour to the Hebrides and Johnson's diary of a journey into north Wales.* 6 vols. Oxford.
Hird, S., and Peeters, M., (1991). UK protection for recombinant DNA – exploring the options. *European Intellectual Property Review*, **9**, 334-339.
Horton, A., (1994). European design law and the spare parts dilemma: the proposed regulation and directive. *European Intellectual Property Review*, **2**, 51-57.
Hurt, R. M., and Schuchman, R. M., (1965). The economic rationale of copyright. *American Economic Review*, **56**, 421-432.
Huygen, F., (1989). *British design: image and identity.* London:
Institute of Trade Mark Agents (1994). *UK Trade Marks Act 1994: a practical guide.* London: Longman.
Jacob, R., and Alexander, D., (1993). *A guidebook to intellectual property: patents, trade marks, copyright and designs.* 4th edition. London: Sweet & Maxwell.
Johnston Report (1962). *Report of the departmental committee on industrial designs.* Cmnd 1808. London: HMSO.
Kelly, F., (1988). *A guide to early Irish law.* Dublin.
Korah, V., (1972). Dividing the common market through national industrial property rights. *Modern Law Review*, **35**, 634.
Laddie, H., Prescott, P., and Vitoria, M., (1995). *The modern law of copyright and designs.* 2nd edition. London: Butterworths.
Landes, W., and Posner, R., (1989). An economic analysis of copyright law. *Journal of Legal Studies*, **18**, 325-363.
Macaulay, T. B., (1853). *Speeches parliamentary and miscellaneous.* 2 vols. London.
McCarthy, F., (1979). *A history of British design 1830-1970.* London.

MacLaren, J. (1870). G J Bell, *Commentaries on the laws of Scotland and on the principles of mercantile jurisprudence.* 7th ed. Edinburgh.
Macmillan, D., (1986). *Painting in Scotland: the golden age.* Oxford: Phaidon Press.
Macmillan, D., (1990). *Scottish art 1460-1990.* Edinburgh: Mainstream.
MacQueen, H. L., (1989). *Copyright, competition and industrial design.* 1st ed. Aberdeen: Aberdeen University Press for The David Hume Institute.
MacQueen, H. L., (1994). Extending intellectual property: producers v users. *Northern Ireland Legal Quarterly*, **45**(1), 30-45.
MacQueen, H. L., Lloyd, I., Henderson, H., and Tyre, C., (1993). Intellectual property. *The laws of Scotland: Stair memorial encyclopedia*, **18**, paras 801-1664. Edinburgh: Butterworths.
MacQueen, H. L., and Peacock, A., (1995). Implementing performing rights. *Journal of Cultural Economics*, forthcoming.
Max Planck Institute (Munich) (1991). Towards a European design law. *International Review of Industrial Property and Copyright*, **22**, 523
Meek, R. L., Raphael, D. D., and Stein, P. G., (1978). Adam Smith, *Lectures on jurisprudence.* Oxford: Oxford University Press.
Miller, C. G., (1994). Magill: time to abandon the 'specific subject-matter' concept. *European Intellectual Property Review*, **10**, 415-421.
Millgate, J., (1986). *Sir Walter Scott's magnum opus and the Pforzheimer manuscripts.* Edinburgh: National Library of Scotland.
Millgate, J., (1987). *Scott's last edition: a study in publishing history.* Edinburgh.
Monopolies Commission (1985a). *Ford Motor Company Limited.* Cmnd 9437. London: HMSO.
Monopolies Commission (1985b). *The British Broadcasting Corporation and Independent Television Publications Ltd.* Cmnd 9614. London: HMSO.
Monopolies Commission (1988). *Collective licensing.* Cm 530. London: HMSO.
Monopolies Commission (1994). *The supply of recorded music.* Cm 2599. London: HMSO.
Morcom, C., (1994). *A guide to the Trade Marks Act 1994.* London: Butterworths.
Morris, A. I., and Quest, B., (1987). *Design: the modern law and practice.* London: Butterworths.
Mossner, E. C., (1980). *The life of David Hume.* 2nd ed. Oxford: Clarendon Press.
Mossner, E. C., and Ransom, H., (1950). Hume and the "conspiracy of the booksellers": the publication and early fortunes of the *History of England. University of Texas Studies in English*, **29**, 162-182.
Nicholson Report (1983). *Intellectual property and innovation.* Cmnd 9117. London: HMSO.
Office of Fair Trading (1986). *Review of restrictions on the patent agents' profession.* London: HMSO.
Palmer, T., (1990). Are patents and copyrights morally justified? The philosophy of property rights and ideal objects. *Harvard Journal of Law and Public Policy*, **13**, 817-865.
Paton, G. C. H., (1940-58). *Baron David Hume's lectures 1786-1822.* 6 vols. Edinburgh: The Stair Society.
*Parliamentary Debates.* London: HMSO.
*Parliamentary History* (1813). *The parliamentary history of England from the earliest period to the year 1803, volume XVII, AD 1771-1774.* London.
Patent Office (1994). *Annual report and accounts 1993-94.* London: HMSO.
Pevsner, N., (1975). *Pioneers of modern design.* 2nd revised ed. London: Pelican.
Phillips, J., (1986). *Introduction to intellectual property law.* 1st ed (2nd and 3rd eds with A Firth in 1990 and 1995). London: Butterworths.

Phillips, J., (1993). International design protection: who needs it? *European Intellectual Property Review*, **12**, 431-436.

Plant, A., (1934). The economic aspect of copyright in books. *Economica*, **1** (ns), 167-195.

Plant, A., (1953). *The new commerce in ideas and intellectual property*. Stamp Memorial Lecture, University of London. London: Athlone Press.

Price, T., (1993). *The economic importance of copyright*. London: Common Law Institute of Intellectual Property.

Puri, K., (1990). The term of copyright protection: is it too long in the wake of new technologies? *European Intellectual Property Review*, **1**, 12-19.

Reichman, J. H., (1994). Legal hybrids between the patent and copyright paradigms. *Columbia Law Review*, **94**, 2434-2558.

Reynolds, J., and Brownlow, P., (1994). Increased legal protection for schematic designs in the United Kingdom. *European Intellectual Property Review*, **9**, 398-400.

Ricketson, S., (1987). *The Berne Convention for the protection of literary and artistic works 1886-1986*. London: Queen Mary College.

Rose, M., (1993). *Authors and owners: the invention of copyright*. Cambridge, Mass: Harvard University Press.

Ross, I. S., (1972). *Lord Kames and the Scotland of his day*. Oxford: Clarendon Press.

Sherman, B., and Strowel, A., (1994). *Of authors and origins: essays on copyright law*. Oxford: Clarendon Press.

Skilbeck, J., (1988). *The export performance of the copyright-dependent industries*. London: Common Law Institute of Intellectual Property.

Staines, I. A., (1983). Protection of intellectual property rights: Anton Piller orders. *Modern Law Review*, **46**, 274.

Stewart, S. M., (1989). *International copyright and neighbouring rights*. 2nd ed. London: Butterworths.

Tompson, R. S., (1992). Scottish judges and the birth of copyright. *Juridical Review*, **37** (ns), 18-42.

Todd, W. B., (1976). Adam Smith, *An inquiry into the nature and causes of the wealth of nations*. 2 vols. Oxford: Oxford University Press.

Tootal, C., (1990). *The law of industrial design*. London: CCH Editions.

Trade and Industry Select Committee (1986-87). *The UK motor components industry*. HC 407. London: HMSO.

Trebilcock, M. J., (1976). The doctrine of inequality of bargaining power: post-Benthamite economics in the House of Lords. *University of Toronto Law Journal*, **26**, 359.

Turner, B., (1993). A true design right: *C & H Engineering v F Klucznik & Sons*. *European Intellectual Property Review*, **1**, 24-25.

Tyerman, B. W., (1971). The economic rationale for copyright protection for published books: a reply to professor Breyer. *University of California at Los Angeles Law Review*, **18**, 1100-1125.

Whitford Report (1977). *Copyright and designs law: report of the committee to consider the law on copyright and designs*. Cmnd 6732. London: HMSO.

World Intellectual Property Organisation (1995). *Committee of experts on the development of the Hague agreement concerning the international deposit of indutrial designs: fifth session, Geneva, June 13-16, 1995*. H/CE/V/2. Geneva: World Intellectual Property Organisation.

# Abbreviations

| | |
|---|---|
| AC | Appeal Cases (House of Lords and Judicial Committee of the Privy Council) |
| All ER | All England Reports |
| Bro PC | Brown's Parliamentary Cases |
| Burr | Burrell's Reports |
| Ch | Chancery Division Reports |
| CMLR | Common Market Law Reports |
| D | Dunlop's Reports of Cases decided in the Court of Session, etc |
| ECR | European Court Reports |
| EIPR | European Intellectual Property Review |
| FSR | Fleet Street Reports of Industrial Property Cases from the Commonwealth and Europe |
| GWD | Green's Weekly Digest |
| M | Macpherson's Reports of Cases decided in the Court of Session, etc |
| Mor | Morison's Dictionary of Decisions in the Court of Session |
| QB | Queen's Bench Reports |
| R | Rettie's Reports of Cases decided in the Court of Session, etc |
| RPC | Reports of Patents Design and Trade Mark Cases |
| SC | Session Cases |
| SLT | Scots Law Times |
| WLR | Weekly Law Reports |

# Index

Berne Convention on the Protection of Literary and Artistic Works (1886), 10, 21, 55, 58, 61, 66, 101

Copyright
  and competition law, 17–23, 40–5, 47, 101, 104
  artistic craftsmanship, 33, 39, 56, 65–7
  compulsory licences, 21–2
  computer programs, 13, 39, 102
  criminal offences, 11–12
  economic analysis of, 12–15
  exhaustion of rights, 19–20
  history of, 1–9
  in designs, 23–4, 31, 32–48, 50, 61–2, 64, 65-7, 77, 81, 88, 90, 95, 98–9, 101–2
  infringement of, 9–10, 11, 12–13, 36–8, 40, 61, 64, 65, 71, 81–4, 93–4
  international framework of, 10–11, 32
  licence to repair and, 45–6
  licensing, 2–3, 16, 20–3, 36, 41–2
  no derogation from grant and, 46–7, 64
  originality test, 8, 57, 64, 72–3, 85
  public interest and, 15–23, 64, 88, 102
  publishing and recording contracts, 93
  specific subject matter of, 19–23
  term of, 13, 14, 20, 32, 33, 34–6, 57–9, 66, 99

Design rights
  and biotechnology, 71, 72
  and character merchandising, 66–7
  and circuit designs, 38, 70–1, 99
  and public interest, 88–90, 94–5
  and tactility, 99
  and TRIPS, 100–1
  *See also* Registered designs, Unregistered designs

Franchise agreements, block exemption, 44

Hague Convention concerning the International Deposit of Industrial Designs (1925), 61, 87

Monopolies Commission, 17–18, 20, 40–2, 47, 49, 58, 63, 86, 90, 91, 92, 93, 104

Paris Convention on the Protection of Industrial Property (1883), 32, 61

Passing off, 27

Patents, 25–6, 31–2, 60, 63, 71, 75, 80, 85–6, 88, 92, 103, 104
  Community, 19
  design, 62
  European, 25–6, 32
  petty, 62

Registered designs, 28–31, 33, 34, 39–40, 43–4, 54–6, 58, 60, 61–2, 63, 64, 66, 67–70, 85–6, 87, 90–1, 94–5, 96
  article, applied to, 69, 91, 95
  Community, 87, 95–100; and spare parts exceptions, 99–100
  eye appeal, 28–31, 68–9, 80, 91, 96
  functionality, 28–31, 68, 91, 96

novelty, 28, 30–1, 57, 85, 95
  term, 28, 57–9, 67–8, 80, 90–1, 92, 96
  *See also* Copyright in designs,
  Design rights, Unregistered
  designs.
Spare parts, 17, 20–1, 35–6, 40–8,
  50–1, 52, 53, 54, 63–4, 69–70,
  73–4, 75–7, 86, 91–4, 99–100, 104.
Topography rights, 10, 39, 50, 56,
  57, 60, 63, 71, 74, 81, 96
Trade and Industry Select
  Committee Report on the Motor
  Components Industry (1986–87),
  52–4, 85, 95
Trade marks, 18, 26–8, 31–2, 44, 90
  Community, 19, 28, 44
  and shapes of products, 27–8
  *See also* Passing off.
Trade secrets, 88, 103
TRIPS, 10–11, 87, 100–1
Unfair competition, 62, 94, 98, 104
Unregistered designs, 55, 56, 57, 59,
  60, 64, 70–86, 87, 90, 91–2, 95
  commonplace, 71–3

Community, 87, 95–100; and
  spare parts exceptions, 99–100
compulsory licence, 59–60, 85–6
Crown rights, 60, 86
design documents, 65, 66–7, 77,
  78, 81–3
functionality, 77, 91
infringement, 78, 81–4, 86, 97;
  remedies, 84–5, 86
licence of right, 59–60, 79–80, 85,
  86, 91–2, 100, 103
must fit exception, 51, 53, 73–7,
  92–4
must match exception, 53, 73–7,
  91, 92–4
originality, 71–3
ownership, 78–9
reciprocity, 85
shape or configuration, 70–1
term, 79, 86, 91, 92, 97, 100, 103
topography of semiconductor
  chips, *see* Topography rights
Utility models: *see* Patents, design;
  Patents, petty.

## HOW TO ORDER

### Subscription Rates

**Volume 3, 1995**
**ISSN:** 1350-7516

Four issues:
January, March,
July, November.

**Institutions:**
UK and EEC  £68
Overseas     £76
N. America   $116

**Individuals:**
UK and EEC  £34
Overseas     £38
N. America   $58

**Back Issues/ Special Issues**
£9.95/ $18

**Postage:**
Surface postage is included in the subscription. Please add £10 or $18.00 for airmail delivery.

**Return this form with your payment to:**

Kathryn MacLean
Journals Dept
Edinburgh University Press
22 George Square
Edinburgh EH8 9LF

☐ Please enter my subscription to Hume Papers on Public Policy, Volume 3, 1995

☐ I enclose a cheque (made payable to Edinburgh University Press Ltd)

☐ Please debit my Visa/Mastercard Account number:

_____

Expiry date_____

Name_____

Address_____

_____

_____

Postcode_____

Country_____